ENCYCLOPAEDIA OF
NATURE OF GEOGRAPHY

ENCYCLOPAEDIA
OF
NATURE OF GEOGRAPHY

(Volume 1)

A.Z. Bukhari

ANMOL PUBLICATIONS PVT. LTD.
NEW DELHI - 110 002 (INDIA)

ANMOL PUBLICATIONS PVT. LTD.
4374/4B, Ansari Road, Daryaganj
New Delhi - 110 002
Ph.: 23261597, 23278000
Visit us at: www.anmolpublications.com

Encyclopaedia of Nature of Geography

First Published, 2005

ISBN 81-261-2443-1 (Set)

PRINTED IN INDIA

Published by J.L. Kumar for Anmol Publications Pvt. Ltd., New Delhi and Printed at Mehra Offset Press, Delhi.

Contents

Preface

The universe, which we live in is full of geographical dimensions. Our world has incorporated in itself, varied social, cultural, political, religious and economical aspects of specific significance. Following such wonderful features, natural wealth and the economic progress made, some countries are termed as developed, while some as developing and the rest as underdeveloped. However, all nations of the world, despite all the odds have been interacting with each other for millennia. They have cooperated, and are still co-operating and would continue to cooperate with each other in future also.

Undoubtedly, the world has diverse and dissimilar physical features and people have different complexions, different cultures and different ways of living. It is only due to different geographical situations, but we share all these with each other in many ways. We should thank God Almighty for giving us such beautiful natural gifts, differing widely from each other, to be able to fulfil our diverse kinds of basic requirements. Due to geographical variations, we find at some places, abundance of coal and at others, abundance of gold and so on. Our geographical conditions have compelled us to go on interacting with each other and also being dependent on each other.

Different natures of soils, plateaus, mountains , islands, icelands, flora, fauna, water resources, river systems, landscapes, climate, monsoon etc., are our geographical assets and we are fortunate enough that we have such kind of possessions.

This beautiful world of ours has diverse cultures, religions, languages, and people of different castes and communities and different physical features as well as complexions. Still we ensure unity and integrity in diversity, solemnly. Almost all the countries in the world have got their own political systems and are developing in their respective ways.

Geography is a vast subject and a perfect discipline in its own right. It has various streams and branches. Studying Geography is not an easy task in the modern context. Each stream demands separate books on selected topics. Hence, there is still a dearth for good books. This endeavour is in the same direction. Here, we offer a series of books on all branches and streams of the discipline — Geography. This encyclopaedic study is a milestone in the area of the subject.

The Editor is confident that he has accomplished a good job, for which there has been a room for a long time. Hopefully, this courageous effort of his would be accorded a warm welcome by all concerned — scholars, researchers, students and professionals, alike. And that achievement is not small.

Editor

1

Concept and Perception

Modern Geography

Although the roots of geography, as a field of study, reach back to Classical Antiquity," its establishment as a modern science was essentially the work of the century from 1750 to 1850. The second half of this period, the time of Humboldt and Ritter, is commonly spoken of as the "classical period" of geography. Undoubtedly the extraordinary accomplishment of each of these men, working at the same time but in very different ways, and the influence of their work on all subsequent geography justifies our regarding them as the first masters of modern geography-in that sense as the "founders." But in applying such titles one is easily led to ignore the importance of previous workers-those, we might say, who laid the foundations for the founders, *"Carl Ritter hat auch seine Vorganger,"* Peschel is quoted as saying, and the statement applies as well to Humboldt. Among these predecessors we are not here concerned with the geographers of Classical Antiquity nor with the majority of the writers of geographical works during the two or three centuries preceding 1800.

Our interest is with that smaller, but still considerable, number of students of the second half of the eighteenth century who consciously strove to convert a more or less miscellaneous and useful study into an independent science. Although one could hardly affirm that they had accomplished that purpose, nevertheless it is dear that they had put down the main outlines of the science of geography, as we know it today, before the appearance of geographical publications by Ritter and Humboldt at the turn of the century. They had not *established* those outlines, however. It is for that work, the establishment of the modern science of geography, that we are indebted to Ritter and Humboldt. Nevertheless, most of the fundamental concepts of geography including almost all of those which we will have to examine in this paper-may be found in the writings of the German geographers immediately preceding Ritter and Humboldt. It hardly seems too much to say that, had neither of these men lived, the development of geography after 1800' would have led ultimately, even though far more slowly, to something like that which we know.

As the inheritors of the geography of Classical Antiquity and that of the Renaissance, the geographers of the eighteenth century as a matter of course considered all kinds of phenomena whose differences in different parts of the world made them appear significant to a knowledge of the world. They echoed the famous statement of Strabo: "Geography, in addition to its vast importance to social life and the art of government. . . acquaints us with the occupants of the land and ocean and the vegetation, fruits, and peculiarities of the various quarters of the earth, a knowledge of which marks him who cultivates it as a man earnest in the great problem of life and happiness". The "Physical Geography'" of Immanuel Kant was concerned with all the elements included by Strabo; on the basis of this physical geography, Kant would found "other geographies" including political, commercial, moral (in the sense 'Of *mores)*, and theological geography, as well as the descriptions of each particular land. Physical geography is the essential propaedeutic for an understanding of our perceptions of the world, whether those received directly by travel or indirectly by reading. In brief, "it serves as a suitable arrangement for our perceptions, contributes to our intellectual pleasure, and provides rich material for social discourse".

These statements of Strabo and Kant are susceptible of translation in terms of modern science. What Leighly refers to as "the unspecialized curiosity in the minds of stay-at home readers" was not considered by Kant as in opposition to science. On the contrary the curiosity which desires a knowledge of the world is the stimulus that requires science to pursue that knowledge and seek to organize it. For this purpose, to be sure, it is necessary that individual students and groups of students should specialize, but Kant recognized that there is more than one form of specialization. While he recognized the importance of the specialization according to kinds of objects (which was later to dominate the field of the natural sciences), he saw that form of specialization was not possible in geography, since its single object, the earth, is for us unique in the universe. We shall consider later the form of specialization which he outlined for geography.

The great majority of the writers of geographical works of the seventeenth and eighteenth centuries, however, did not attempt to translate the purpose of geography in terms of scientific interest. Rather, they were concerned with its practical utility. Over and over again they emphasized the value of the study of geography as a means for other purposes-for an understanding of history and as a practical aid to government. Wisotzki has shown how this utilitarian point of view prevented scientific progress in geography.

Overlooking earlier but unsuccessful efforts, we need to consider a small but increasing number of writers in the-second half of the eighteenth century who worked to elevate geography from its subordinate position to that of an independent science, *"reine Geographie,"* or "pure geography," as they came to call it. In contrast with later uses of this term, both in Europe and America, the reader will note the suitability of its use in this connection.

This movement found its first expression in the demand that geography should consider its object, the world, in terms not of political divisions but of more real and lasting divisions of nature. Divisions of area can be established only by boundaries, so that this concept found its first expression in a demand that geography should divide the world by "natural boundaries" of "lands" rather than by political boundaries of states. This movement, which represented an echo from Strabo, was expressed by many writers, both geographers and students of law (notably Grotius), as early as the 16th century. It could make little headway, however, so long as geography was considered to be merely the handmaid of history and government. Although it was argued that the frequent changes in political boundaries made the geographer's work seem of but transient value, the argument was irrelevant so long as it was the major business of the geography teacher to provide information about political areas. To accomplish this task at a time when the political map of central Europe-Germany and Italy-was a crazy-quilt of hundreds of states, the geography teacher could hardly spare time to consider "natural regions."

On the other hand it seemed obvious to those who were attempting to develop a scientific geography, that the conventional organization of the knowledge of different parts of the world by political divisions was anything but satisfactory. Whether in terms of the political map of Europe of that day or this, or of the political map of the United States today, it hardly seems too much to say that no student who is seriously concerned with the attempt to develop regional geography could regard political divisions as even "generally satisfactory."

Then as now, however, the effort to consider the areas of the world in divisions other than political presented the difficult problem of finding all equally definite method of dividing areas of the lands. The answer to this long-felt need appeared to be offered by a theory put forward by various students in the middle of the eighteenth century, notably Buache in France—namely, the theory of the continuous network of mountain systems see Wisotzki's chapter on *"Der Zusammenhang der Gebirge,"*. The theory was immediately taken up in Germany, and Gatterer made it the basis of a physical division of the world into lands and regions *(Gebiete)*. Whereas the similar attempts made much earlier-in Holland and Italy as well as in Germany-had been largely ignored, Gatterer's work had a marked influence on his contemporaries and on later students, including Humboldt.

We may therefore mark the beginning of a continuous reformation of geography with the appearance of his publications in 1773-75.

A somewhat different stimulus to the development of scientific geography throughout this period came from the lecture course on "physical geography" that the philosopher Immanuel Kant presented at the University of Konigsberg, repeated some forty-eight semesters during the period 1756 to 1796. For Kant himself, as Gerland has noted, the study of geography represented only an approach to empirical knowledge necessary for his philosophical considerations. But finding the subject inadequately developed and organized, he devoted a great deal of attention to the assembly and organization of materials from a wide variety of sources, and also to the consideration of a number of specific problems-for example, the deflection of wind direction resulting from the earth's rotation.

It is difficult to estimate the historical importance of Kant's work in our field. His course in physical geography, one of the most popular of his courses, was heard by a very large number of students during its forty-year period, and long before the publication of the lectures, after 1800, there were many fairly complete handwritten copies in circulation. Furthermore, former students, like Herder, had given wider circulation to his ideas. Whether by indirect means, or only after their publication in 1801-05, the lectures had a significant influence, as we shall see, on both Humboldt and Ritter.

The greater part of Kant's course followed the outline of "general geography" that Varenius had laid down a century before and which Lulofs had followed much later (in German translation, 1755). In addition to the large amount of material taken from Lulofs, Kant later drew from the work of Bergman (published in Swedish, 1766, in German, 1769), as well as Buache, Buffon, and especially Busching. In his relatively briefer consideration of individual countries, however, he does not appear to have been stimulated either by Buache or Buffon to have altered Busching's conventional division by political units; certainly the outline and material of the course were well established before the appearance of Gatterer's work.

Nature of Field

We cannot here consider the various steps in the subsequent development of the "pre-classical" period, and can only briefly suggest the underlying factors that stimulated its progress of great importance was the rapid growth in various physical sciences. In particular the development of reasonably accurate barometric methods, notably by de Luc, whose work was first published in 1772, made it possible to measure the "vertical" dimension of the earth's surface. The fundamental approach of many geographers of this and the following period was notably influenced by the great reformer of education of the time, Pestalozzi, of Zurich, who brought into German schools the philosophical viewpoint of Rousseau. Closely related to the change in geographic thought-whether as cause or effect-was the so-called "natural division" of France into departments according to rivers and mountains. Somewhat later the radical and repeated changes in political boundaries during the Napoleonic Wars helped to discredit geography texts based on such fluctuating conditions, and thus helped to make way for a geography based on the physical description of the world.

Our present interest is in the nature of the field of geography as it was conceived by a small but active minority of writers in the late eighteenth century who, before Humboldt and Ritter, were endeavouring to establish geography as a science. We need not attempt to set a definite limit as the end of this period, but may consider it as overlapping perhaps a decade or so into the nineteenth century. Although Humboldt published at least one geographical article before his all-important journey to tropical America in 1799, his major publications in our field did not begin until after his return in 1804, the year in which Ritter's name first appears in geographical literature. According to Marthe, however, it was not until nearly 1820 that each of these two figures assumed a dominant position in geography.

The geography which the late eighteenth century had inherited from the past was essentially limited to phenomena of the world as known to man, that is, of the earth's surface, in the broad sense of a surface of some thickness extending into the atmosphere and into the solid earth wherever man had penetrated. Kant had described the field of physical geography as the world *(Welt)*, so far as we can come in relation to-it, the scene of our experiences. The study of the earth body belongs to mathematical geography but is considered in physical geography insofar as it causes differences in its different parts. In most of the geographies of this period the treatment of the earth-body was confined to an introductory chapter—*Mathematische Vorbegriffe,"* according to Kant and the actual study of geography confined to the earth-surface: The specific term" *Erdober-flache"* was frequently used, and in 1820 Wilhelmi specifically limited the field of geography to the earth-surface.

The geography of this period included both "general" studies of particular kinds of phenomena of the earth surface, and descriptions of many kinds of phenomena found in particular areas. The inclusion of these two form of study within the single field has no doubt been the cause of more controversy than any other single problem in the methodology of geography. It is significant therefore that it was not introduced into modern geography as a result of the chance combination of Humboldt and Ritter. On the contrary the same difference is found in the work of the geographers of antiquity, as Hettner observed in his first brief treatment of the history of geography. The tendency of these two directions to come in conflict with each other and to interchange positions within geography at different times is evidence, he suggests, that they do not represent separate sciences but merely different directions within the same science. The distinction between the two directions was perhaps first clearly stated by Varenius (a German, Bernard Varen, living in Amsterdam) in 1650. In contrast with most of the writers of his time who limited geography largely to a bare description of the several countries-including therein much of their political constitutions-Varenius divided the field into "general or universal geography" and "special geography" or "choreography note Humboldt's discussion of Varenius. His untimely death, at the age of twenty-eight, in 1650 prevented him from studying the second part, though his previous work on the geography and history of Japan would indicate that he did not lack interest in that part of the field.

The terms which Varenius had used, "general" and "special" geography, later became the standard terms in Europe for these two aspects of the field, though many later writers have found them unsatisfactory. As we will note in a later section, the frequent use by German writers of the adjective *"systematisch"* in describing "general geography" supports the common use in this country of the term "systematic geography." The term "special geography" was largely replaced in German literature by *"Landerkunde;'* which in spite of obvious disadvantages is commonly favoured over non-Germanic terms, "special," or-the term now nearly universal outside of Germany—regional geography.

Following Varenius, Lulofs and Bergman had concentrated on systematic geography. Kant, however, though depending in part on them, likewise suggested the place of *Landerkunde,* but his beginning of such a regional study is not firmly based on his systematic studies. The distinction between the two aspects of geography was more explicitly stated by Gatterer, 1773-75, by Krug, 1800, and particularly by Bucher, 1812.

Systematic Studies

Although the systematic studies of this period undoubtedly were of major importance in the progress of geography toward scientific standing, Plewe finds that many of them, notably those of Bergman, tended to fallout of geography and become studies in other sciences. In contrast, John Reinhold Forster-the first of the great scientific geographic travellers-consistently maintained the "macroscopic" sense suitable to geography and endeavoured to explain the regional relations of each type of phenomena studied. The work of S.R. Forster, and particularly that of his son, George, was regarded as of great importance by Humboldt, upon whom, as we shall see, they had a determining effect. Plewe, among others, counts the elder Forster as the first—in order of time-of the great "classical geographers".

The division between general or systematic geography and *Landerkunde,* or regional geography, represented therefore a form of dualism that was characteristic of geography throughout its initial period of development as a modern science. This form of dualism in geography is not to be confused, as is often the case, with a dualism in terms of content, as Wagner warned, in 1890.

In terms of content, the eighteenth century writers commonly distinguished three divisions: "mathematical geography," "physical geography," and what was variously called "historical" or "political geography," but which is not to be confused with the meanings of those terms as used-today (or as used by Kant).. Mathematical geography consisted in large part of the study of the earth as an astronomical body. As we have seen, it was always of minor importance and in practice its study was left largely to astronomer. Zeune, in 1808, and Butte, in 1811, definitely excluded it from their outlines of geography.

Major Contrasts

The major contrast in content was that between what was included in physical geography and that contained in what we might calls social geography. With the much more rapid increase in development of scientific methods of studying physical, as distinct from social, phenomena the contrast between the two became more marked. Attempts, to develop a unified science included suggestions of confining the field to what was then called "physical geography." Hut if the present-day student considers this period only in terms of the titles which its writers used, he will have an erroneous conception of their works. As used by nearly all the writers of this period including Kant, Forster, and later, Humboldt-the term "physical geography" was not limited to that which later was to be called the "physical" or "natural'; environment, but included races of men and, commonly, their physical works on the earth. Thus Kant not only included man as one of the features "encompassed in the earth surface *(Erdboden)"* but also considered man as one of the five principal agents affecting changes on the earth. Indeed, Kant's "physical geography/, both in purpose and in content, might be considered as "anthropocentric," a point of view which Ritter inherited from Kant (according to Becker).

Very few writers of this period -made 'the distinction which is most familiar to us, between a geography of natural, or non-human features, and a human geography, though it may be found in the work of J. M. F. Schulze, 1787, and of Ruhle, 1811. For most writers of this and the following period.—*i.e*, until perhaps the middle of the nineteenth century—the term "nature" *(Natur)* was not used in contradistinction to human, but to indicate that which was perceived externally in contrast to one's Internal feelings and thoughts. So far as I can find, no geographer of this period questioned the place of either human or organic features of the earth in the field of geography. One might note, however, a trend in that direction, in the interpretation which Hommeyer, in 1810—and perhaps others-gave to *"reine Geographie,"* namely, as limited to the conditions of the terrain; the study of climates, minerals, and organic life was left to *"Naturkunde," "Natur-beschreibung"* and *"Landerkunde"*.

Although the writers of this period recognized that the scientific geography which they were attempting to develop included somewhat separate parts—"physical" and social geography-they considered these to be associated not merely because the phenomena were to be found in the same places, but because they were causally connected by mutual relationships. No doubt the interest in relationships between different groups of men and the Particular character of their natural environment has ever been considered appropriate in geography. Kant, for example, noted that "in the mountains, men are actively and continuously bold lovers of freedom and their homeland" and similar relations were discussed by Gatterer and his followers. Kant specifically based the social branches of geography on their relations to physical geography: "theological principles in many cases undergo important changes as a result of differences in the land *(Boden)*". In this he was followed by Muller, in 1785, and Fr. Schultz, in 1803, and by many others in the following decade. So far as I can, find, however, it did not occur to, any of the students of this period to consider these relationships as in themselves the direct object of geography study.

During the latter part of this period, it became common to regard the intricate interlacing of relationships between all the phenomena found in the earth surface, organic and inorganic, as functional relations of parts of a single whole. The concept was early expressed by Kant, for whom the systematic arrangement of objects by classes, as in the system of Linnaeus, divided nature into parts without organizing a system, whereas physical geography "gives an idea of the whole, in terms of area". At the beginning of the following century this concept was apparently so strongly supported by the philosophical views of the day that Butte stated, in 1811: "no scientist doubts the reality of an earth organism". Similarly, within any particular area, the combination of all interrelated phenomena is not a mere aggregate but an interrelated "Whole." according to Krause as well as Butte, in 1811.

Few of the writers of that periodor indeed of any later period-distinguish clearly between the concept of unity of all the phenomena at any particular place or area, in what we may call a vertical totality or unity, and the horizontal unity of the area as an individual unit distinct from neighbouring units. This latter concept was, as Bucher noted in 1827, in a sense forced upon the promoters of the new school of geography in competition with the old. In place of the definite natural units of states, sharply defined by political boundaries, the new geography required equally definite "natural" units, somehow defined in nature. For a time, such definite natural boundaries appeared to be provided by drainage basins sharply separated by the "network of mountains," and, of course, by the sea. As increased knowledge of the actual conditions of the earth's surface made this theory untenable" the problem of finding "natural boundaries" for such "natural units" of area became much more difficult. We need not here concern ourselves with the long conflict between the supporters of "dry boundaries" (watersheds) and "wet boundaries" (the rivers), nor with the attempts to combine them. In large part, the general theory seems to have been supported by the concepts of natural philosophy which had a marked influence on nearly all the geographers at the end of this period... Indeed the expression of this philosophical concept in the geography of the time was taken up by political leaders-at least when it suited their particular purposes. On this basis Dalton justified the expansion of revolutionary France to the Rhine Sporer's long footnote on similar political arguments of a later period, 68, 363].

While the concept of natural divisions of the lands was effectively introduced as early as 1773, by Gatterer, the consideration of each of the divisions as in itself a natural unit does not appear clearly until 1805, in the work of Hommeyer. With Hommeyer, as with his predecessors, this unity represented perhaps little more than this unity of the landforms. The concept of a composite unity-the integrated total of all the phenomena of an area into an individual unit, distinct from those or neighbouring areas was stated emphatically by Zeune, 1808, and by Butte in 1811. For Butte the individual lands and districts were "organisms" which like any organism included a physical side-inanimate nature, and a psychical side-animate nature including man. "The unit areas *(Raume)* assimilate their inhabitants" and "the inhabitants strive no less constantly to assimilate their areas".

Other writers of the period, however, did not accept these concepts without question. Ruhle von Lilienstern immediately, 1811, opposed Zeune's theory with the impracticability of establishing definite "natural regions". A decade later both Wilhelmi and Selten urged that the boundaries of unit areas were not to be determined on the basis of anyone kind of phenomena and recognized the difficulty of establishing definite limits where the boundaries of many different kinds were each gradual rather than sharp and, taken together, failed to coincide. "Nevertheless," said Wilhelmi, "the forms of nature are clear and distinctly separated as soon as we regard the particular constitution (of the total of all factors in any area) in its full form, rather than the boundary or transition zone".

The most thoroughgoing criticism of these concepts was made by A. L. Bucher in 1827. In a discussion fifteen years earlier, he had, with some hesitation, adhered to the concept of natural boundaries and apparently also —though this point is not clear to that of regions as natural units. Ruhle's critical discussion of Zeune, however, had made a strong impression upon him, his own efforts to establish 'natural divisions between parts of the continents had made him still more skeptical of the theory, and the discussions of Selten and particular Wilhelmi revealed, he thought, its fundamental fallacies—even though neither of them had arrived at that final conclusion of his concern over this and similar problems in geography led him to make a critical examination of nearly forty geography texts published in Germany in the first quarter of the nineteenth century. This remarkable study, whose value was at once perceived by Berghaus, merits a high rank among critical works in geographic methodology. Wisotzki rightly gives its discussion a prominent place at the conclusion of his survey of the pre-classical period. The arguments with which Bucher attacked the concepts of natural boundaries and natural regions leave little for the present-day writer to add. Indeed it would hardly seem too much to say that if Bucher's study had not been so generally overlooked by recent German as well as American geographers, much of the discussion of regions as natural concrete units in a later section of this paper would be superfluous. As the culminating expression of methodology of regional geography of the pre-classical period, Bucher's study represents in a sense a dead end. The long search for the true "natural divisions" he convicted as futile from the start, but he would not return to the use of political divisions. Rather he suggested that it was not necessary to establish either boundaries or divisions; regional studies were needed only for special purposes for which the areas concerned could be arbitrarily bounded in any convenient way. The essential work in geography was concerned with the systematic studies of individual categories of phenomena, each however in relation to the earth. As Bucher's final study falls well within the "classical period" we will have occasion to consider these conclusions again in that connection.

Other Features : Certain other features of the geography of the pre-classical period may be more briefly noted. The impression which the view of an area makes upon the spirit of the beholder was emphasized by Hommever who considered the study of the aesthetic-geographic character of the *Landschaft* as a part of geography. Hommever, however, was one of the first to introduce confusion in geography regarding this term, as he specifically defined it as a portion of territory, intermediate in size between a *Gegend* and a *Land*. More important in the development of the aesthetic description of nature, according to Humboldt, were Rousseau, Buffon, Bernardin de St. Pierre (in *Paul et Virginie)*, Chateaubriand, Playfair and Goethe, and, among geographers, particularly George Forster.

It was a natural consequence of the manner of development of the new science of geography that its proponents should see it closely related to history, but they insisted that geography should not be "the handmaid of history," but rather that the two were "sister subjects." The comparison of the two fields was most clearly stated by Kant, as we shall see it a later section. The extent to which geographical studies should include historical treatment was discussed by Ruhle, in 1811, and, notably, by Bucher in 1827.

Though the new geography found its basis in physical rather than political features, it did not exclude the specific problems of political geography, as we understand the term. On the contrary, one of its claims was that it provided a firmer foundation for the study of that field than had a geography which constituted little more than political and historical geography. In particular, the drastic changes in the areas of states during this period constantly engaged the attention of geographers, and as we have noted, the concept of "natural boundaries," first developed as a framework for physical geography, was carried over-all too simply-both into political geography and political practice. One by product of this was the further confusion of the term "natural boundary" to include linguistic boundaries a concept which recently has again been emphasized-not without inconsistency, to be sure-by one of the rulers of Europe.

Although the new science of geography sought its foundations in the more permanent physical features of the earth, it was characteristic of its time in considering these primarily in relation to man. Throughout the period most writers, consciously or unconsciously, followed Kant in regarding physical features in terms of their ultimate importance to man. Muller in 1785, and Kayser, 'in 1810 specifically emphasized the study of the earth as "the dwelling place *(Wohnsitz)* of man." On the other hand, one can observe a swing away from this viewpoint in those writers like Hommever who emphasized the description of the earth forms-in terms of pure description, without explanation-as the sole content of *"reine Geographie"* all other features might be considered in applied geography.

One other concept of geography expressed at this time was that it was a study concerned with the "Where" of things. Stated by Lindner in 1806, it was vigorously opposed by Ritter in the same year.

Finally, it may be added, the students who were consciously endeavouring to construct a science of geography, repeatedly emphasized the importance of two cardinal principles of any scientific work. In contrast with the blind dependence on authority, characteristic of earlier periods, and the *a priori* construction of systems of supposed facts about the earth that characterized much of the work of the eighteenth century, many students toward the end of this period emphasized the importance of determining first the actual facts. Likewise they wrote repeatedly of the need to indicate specifically the sources of information.

In large part, however, the new science of *"reine Geographie"* represented the expression of purposes and hopes rather than of accomplishment. In spite of the insistence on establishing facts, the geographers of this period, with few exceptions, continued to set up *a priori* systems of facts without putting them to the test, and many of these concepts we still inherit—or have constructed anew. In the century and a quarter since Butte spoke of areas as organisms no geographer, so far as I can find, has seriously- attempted to establish the organic character of any single regions and yet, as we will see, current textbooks of the most modern type repeat this *a priori* assumption. In the century since Bucher's scientific integrity led him to announce his failure to establish a division of lands into natural regions, no one, so far as I can find, has seriously attempted to show the 'error of his methods or argument, or to produce what he found impossible; nevertheless the concept of the region as a definite unitary entity is prominent in the writings of a large number of geographers today.

Classical Period

If the geographers of the late eighteenth century developed the greater part of the theoretical concepts of the new science of geography, the transformation of their "ideas, demands, and wishes into facts" was largely the work of Alexander von Humboldt and Carl Ritter. It is an extraordinary fad that modern geography in all lands should owe so much to two men living at the same time in the same country for over thirty years in the same city.

This situation was not merely the result of a coincidence. Both of the "founders" of geography depended in large part on their predecessors whom, of course, they had in common. Furthermore, Ritter repeatedly asserted and no student of his work has questioned the tact-that 10 many respects Humboldt, the elder by ten years, was his teacher. Peschel rather emphatically endorses Ritter's statement that, without the work of Humboldt, his own work could never have been produced. Though we may find some exaggeration in the statement, its essential truth is testified by the innumerable places where Ritter echoes the concepts of Humboldt, as well as by his hundreds of specific references to the latter's writings. That these are not merely the meticulous notations of a pedantic scholar is testified by Humboldt himself. In the introduction to his final work on Central Asia, he refers to the earlier publication of a series of itineraries and observes that "the terminology that I proposed for the mountain systems of Asia, and the geologic views on their mutual independence, have been discussed with great superiority by my old and illustrious friend, M. Carl Ritter, in his great work on Asia."

Use of Ritter's Work by Humboldt : Although Humboldt made repeated use of Ritter's work in his own writings-the citations of Ritter in *Asie Centrale* must run well over a hundred-his general concepts of geography were evidently well formed before he met the younger student and before the appearance of Ritter's first geographic writings. Humboldt had been trained neither for geography nor for science, but rather for governmental service—through courses in technology, economics, history, philology, etc. His dominant interest however was in the study of nature—as a mine inspector he was concerned to study subterranean vegetation—and a close relationship with George Forster turned this interest in the direction of geography. In particular a trip which they made together in the Rhineland and to England, in 1790, stimulated Humboldt's ambition to undertake scientific travel and introduced him to Forster's methods of geographic study. It may well be that Humboldt's concept of geography had already been influenced, indirectly, by Kant.

During his short period in governmental service Humboldt published a number of scientific papers, including one or two in geography, but his great work, of course, came as a result of his travels in tropical America, in 1799 to 1804, as well as the later trip to Central Asia, in 1829. It would be an error, however, to suppose that his geographic work was limited to his own observations, or even to the lands which he had observed. Of over sixty years of active geographic study, his explorations hardly totalled five years. (To be sure much of Humboldt's time in the latter part of his life, when he was a very active privy councilor to the king of Prussia, was occupied with state problems.) Many of Humboldt's studies, even before the *Kosmos,* extended over the whole world and depended on the utilization of the work of countless other explorers and scientists. Indeed, during the greater part of his career Humboldt was as much a geographer of the study and library as was Ritter.

Influence of Humboldts on Ritter : Although Humboldt's work had an important influence on Ritter throughout his career, Peschel was unquestionably exaggerating when he stated that their first personal contact, during several weeks in 1807, was the critical experience that definitely turned the younger student to geography. Rattle has shown that neither this statement-frequently repeated by later writers after Peschel-nor the more general inference that Ritter came into geography from history, is supported by the facts of his life. Probably no student of his time-and few since—have been so specifically trained for geography as was Ritter. From very early childhood until he went to the university, he lived in a school, in Schnepfenthal, in which education followed the precepts of Rousseau and Pestalozzi: knowledge of the world was acquired by direct observation of nature on walks and longer trips. Throughout this period the youth was under the particular personal care of J. C. F. GutsMuths a teacher whose major interests were in nature study and in geography, and who himself made some contribution to the development of geography. At this time, geography was Carl Ritter's favourite study, and in the following university years the natural sciences commanded his chief attention together with history and theology. These studies did not end when he left the university: for nearly twenty years as tutor in a private family in Frankfurt, he had both the need and the opportunity to continue the study of natural sciences, geography, and history, and to make numerous field trips in the Rhineland, Switzerland, etc., making careful, detailed observations. Likewise he was able to come into close relations with many eminent scholars of the day, of whom the most significant for him were the anatomist, Sommering, the geologist, Ebel, and the leader of educational reform, Pestalozzi.

Ritter's Preparation for Research : In sum, we may conclude that Ritter's preparation for research covered an extremely wide field, in which on the one hand natural sciences, and particularly the observational methods of nature study, predominated, but in which, on the other hand, his interest was increasingly in human problems, *i.e.,* history. Geography, as he read it in the writings of Herder, a follower of Kant, maintained the connection between his earlier field of studies and- his ultimate interests. Although his meeting with Humboldt, in 1807, undoubtedly stimulated his interest in geography, for the very reason that Humboldt had demonstrated so clearly the importance of earth conditions to man, for at least ten years thereafter Ritter was still uncertain whether to continue his geographic work or to shift to history.

His first publications—overlooking a series of geographic studies for elementary students-dealt with the geography of Europe. These appeared between 1804 and 1807 and made a considerable impression in Germany. One of these, which was also enthusiastically reviewed in France, was a brief study in the systematic geography of Europe, based on six maps, with explanatory text, of individual categories of phenomena. Appearing some years before Humboldt published similar maps, it constituted, according to Plewe, "the first physical atlas". Though all these geographical studies appeared before the celebrated meeting with Humboldt, one may presume that Ritter had already been influenced by the latter's writings, since he wrote at that time that he "had devoured all his published works with-a kind of ravenous appetite".

During the next few years Ritter- prepared a more complete systematic geography of the world, a "Handbuch der physischen Geographie," which he intended to be the completion of his geographic work and a bridge to studies in history, to which he wished then to devote the rest of his life. For some reason however-possibly because of the criticism offered by the geologist, von Buch, who read the manuscript at Ritter's request-he decided not to publish it. Parts of it however were utilized by various colleagues to whom he freely offered it, so that the work appeared in a number of texts of the time, in some cases with the original authorship indicated, in others not.

During the same period Ritter's studies on Asia led him into the historical field, in which he prepared a lengthy study in ancient history (published later, 1820)—according to Rattle "the only work of Ritter of exclusively historical character". Apparently, however, he' contemplated shifting entirely to history. To fulfil his geographical work he outlined a project which should cover the world by region—regionally in a four volume *Erdkunde.* This would then supply the sound foundation for historical studies. Working at the University of Gottingen-where at the same time he took further courses in history, but also in mineralogy he completed the first volume, on Africa, according to plan, but soon found that the enormous amount of material that he collected on Asia could not be handled in one volume, so that he issued the second volume as but a part of the study of Asia. The appearance of these two volumes, in 1817 and 1818, made a tremendous impression in the academic world; almost at once, says Marthe, Ritter was recognized "without ,Question as the reformer of geography, as the master who first made that field into a science".

Arm-Chair Geography of Ritter : Ritter's position as the master of academic geography was therefore definitely established before he had held a single academic post. After one year as professor of history and geography at the gymnasium in Frankfurt, the efforts of various officials in Berlin-including Wilhelm v. Humboldt, the brother of Alexander—were successful in arranging for a double position for Ritter, in the military college and at the university. In 1820, Ritter went to Berlin as the first university professor of geography. At just what time he recognized that his plan of "finishing up on his geographic work in order to go on to' history was unattainable is not clear; in any case he continued to work in geography in Berlin during the remaining 39 years of his life, although it may be said that in the latter volumes of the *Erdkunde,* his interest in history became more and more pronounced.

Following Humboldt's final shift of residence from Paris to Berlin, in 1827, the professional relations between the two increased steadily. According to Kramer, who lived for some years in Ritter's house, few days passed, at least in the latter years, in which-if they were both in Berlin-communications of greater or less importance did not pass between them.

This is not the place to attempt an appraisal of Ritter's contribution to geography, nor should such an attempt be necessary. No geographer in history had his work discussed so repeatedly, and often critically, as Ritter. The studies referred to in the following paragraphs are no mere obeisances to a great "traditional" figure. Except for the quotations from Humboldt, they are taken from critical examinations of Ritter's work by authors who did not hesitate to point out its weaknesses and the extent to which it failed to live up to his stated purposes. Nevertheless those students who have thoroughly examined his work appear to agree unanimously with Humboldt in recognizing its "masterly" character. The more important critical studies of Ritter's work are those of Peschel, Marthe, Ratzel, Hozel, and Wisotzki.

In view of the real differences between Ritter and Humboldt, both in personal character and, as a result, in their work, but particularly because of the supposed opposition between their points of view in geography, it is appropriate to note Humboldt's opinion more fully. The earliest expression that I have found is in the lecture course that Humboldt gave at the University of "Berlin, 1827-28. Humboldt did not limit himself to the laudatory phrases which the occasion called' for when he described Ritter's *Erdkunde*—at that time only the first two volumes had appeared-as "the most inspired work of this kind (comparative geography), that our century has yielded; it is the first work in which is presented the influence which the surface-view has had on the peoples and their fates". In a personal letter to Ritter a few years later, 1832, he expressed his great admiration of the work on Asia, adding that it gave him great pleasure to sing its praises to the king and all at the court, telling them that "such an important work had not appeared in thirty years". Humboldt used his influence at the court to have Ritter relieved of some of his heavy duties, presumably at the military college, in order that he might complete the work on the *Erdkunde*. That Humboldt did not alter his opinion is indicated by numerous statements published in later writings. In the introduction to his Central Asia he finds Guigniaut's verdict correct: "one of the greatest and most magnificent monuments raised to science in our day". In the first volume of the *Kosmos* (1845), he speaks of "Carl Ritter's great and inspired work" which "has shown that the comparative geography attains thoroughness only when the whole mass of facts that have been gathered from various zones, is comprehended in one view, is placed at the disposal of the integrating *(combinierenden)* intelligence". Later in the same volume he refers to Ritter's work as the fulfilment of a part of the plan of Varenius, which-in the earlier period of scientific development. Varenius himself had not been able to accomplish: "It was reserved for our time, to see the *comparative* geography cultivated in masterly fashion, in its widest compass, indeed in its reflex on human history, on the relations of the form of the earth to the direction of the characteristics of peoples and the progress of civilization"

The new direction which Ritter was endeavouring to bring into geography is explicitly stated in many writings from 1804 on. His first principle, frequently repeated, was that geography must be an empirical science, rather than one deduced either from rational principles-from philosophy or from *a priori* theories of "general geography." "The fundamental rule which should assure truth to the whole work," he wrote of his *Erdkunde*, "is to proceed from observation to observation, not from opinion or hypothesis to observation".

The first major plank in Ritter's reform of geography therefore-was to call a halt on all attempts that led from theoretical considerations to systems of earth forms. He was among the first to show that the theory of a continuous network of mountains was opposed by the records of observed facts and that, likewise, there was no general correspondence, as often assumed, between the crest-lines of the mountains and the divides of drainage basins. Although the teleological view of the universe that Ritter had received through Herder from Kant provided the background of purpose of his study, he was not, like Kant, a philosopher which required geography in order to establish theories of world knowledge. Convinced that there were laws governing the relation of human and nonhuman phenomena on the earth, Ritter was in no great hurry to establish them, but rather felt that, if he could bring together all the facts and relationships observed in areas, these would make it possible to state such laws. Attempts to develop these laws first—his own, perhaps, as well as others-had proven unsuccessful: "we must ask the earth itself for its laws".

Ritter was therefore the first great opponent of what may properly be called "arm-chair geography." On the other hand, as a scholar and teacher, rather than explorer or field worker, as Hettner observes, Ritter represented the geographer in library and study: the observations used in his writings were not obtained directly, but rather from the works of others who did observe at first hand. Not that he did not travel and observe, as is often erroneously supposed; on the contrary he travelled much throughout his life, making detailed observations in many parts of Europe. But since he unfortunately became so deeply involved in his studies of Asia as never to commence the volumes on Europe, the detailed observations which he made on his own trips had only the value of enabling him to "understand what others had observed."

In basing his work on observations of other students, Ritter was not satisfied with the limited amount of material which had sufficed for most of his predecessors. In order to achieve the highest degree of accuracy, his plan was to bring together on each point in question the greatest number and variety of trustworthy witnesses, of all times and from all peoples, so that their testimony might be seen together, whether in agreement or for comparison. The material was by no means limited to the works of travellers, whether scientific or otherwise, but was drawn also from the works of a large number of specialized scientists, as one may observe by glancing through the introduction or by noting the footnotes in any of the volumes.

It seems more than probable that Humboldt had Ritter's methods of work, as well as his own, in mind, when he discussed the use of sources in his lecture course given during the winter of 1827-28, at the University of Berlin. "What an individual observer can see, is naturally small in comparison with that which has been observed during so many centuries. If it is therefore very important to gather the observations, it is likewise necessary to be engaged in some part of the sciences of nature, for only so can one learn to understand that which others have observed. For those who can spend the great amount of time and energy necessary, the most important method is the study of all travel descriptions, of all the individual treatises, for only out of the special cases can the general be recognized.

One cannot glance through the pages of Ritter's Presentation of Asia without being impressed by the great number and variety of the sources upon which he depended: the works of contemporary scientists like Buch and Humboldt-whose works are no less honoured by use than by praise—the reports of earlier travellers, whether non-scientific travellers or those of whom Humboldt complained that, while "well prepared in the isolated branches of science," they rarely had "a sufficiently varied knowledge" to study phenomena of different categories in interrelation; and in addition a great host of works of all kinds from the pens of medieval and classical writers. In large degree, these were the same kinds of sources, in many cases the same works, used by Humboldt in his writings. Although the very quantity 'involved made it impossible for Ritter to complete his original purpose of covering the whole world-and indeed deflected him from his particular aim of correlating the human and physical facts established—he at any rate could claim to have assembled for comparison all the evidence available to establish the facts. The first of many to testify to his success in this regard was Humboldt, with particular reference to the area with which he himself was personally familiar, Central Asia. Shortly after Ritter sent Humboldt the first volume on Asia, in its second edition of 1832, Humboldt wrote to him that he could find no words in which "to express the real admiration with which I am filled by your gigantic work on Asia. For the past two years I have been most earnestly engaged with Inner-Asia, with the use of all sources, but how much has become clear to me only in the three days in which, without break, I have been reading in this work. You know everything that has been observed for centuries, with your particular sagacity you arrange everything together, from much-used material you gain ever new and grand views, and you reproduce the whole in the most desirable clarity".

It was not, however, Ritter's purpose merely to assemble a mountainous accumulation of information. The enormous multiplicity of observations must be organized according to the chronological *(raumlichc,* or spatial) principle, which he considered both in a horizontal, and with reference to elevation-in a vertical sense. This did not mean, however, simply an accumulation of facts concerning each area; rather he wished to show the "coherent relation" *(Zusammenhang)* in terms of cause and effect, of the different features' and the "formation of the multiplicity of features into the essential character of an area". By proceeding "from the simple to the compound," he wished to establish the totality of interrelated features as the distinctive character of each area. We are but paraphrasing any number of his statements if we say that he was concerned with reducing to order observations made in nature.

Ritter's Differences : In these respects Ritter's purposes were not so sharply in contrast with those of his predecessors as he believed; but the great difference was in the extent to which he put these into practice—even though this was by no means complete. In certain other respects his position was in clearer opposition. Though he followed the movement of *"reine Geographic"* in using "natural lands" rather than political states, he objected to the overly-simple, and in fact impossible, procedure of dividing only by mountain ridges, or the equally simple division by drainage systems. According to his theoretical statements his divisions were made in terms of relief (in particular he divided the upper, middle and lower portions of strum basins), but in practice he also recognized other factors, as in his separation of the Sahara and in its sub-divisions in terms of climate, vegetation, etc. In this, he was consistent with his fundamental principle, that we should not evolve a theory of natural divisions of the lands, but must always look to nature for its principles.

He reminded his contemporaries, in their intense argument over the relative values of "wet" and "dry" boundaries, that the mere drawing of boundaries was but a means toward the real purpose of geography, the understanding of the content of areas. This view he emphasized by a phrase which, though not clear in itself, was perhaps sufficiently clear in the context in which he used it as to justify its frequent repetition in later decades, so long as the context was not overlooked. Geography, he suggested, is the study of *"der irdisch erfullten Raume der Erdoberflache"*, *i.e.*, the areas of the earth are not to be studied in themselves, as mere divisions of the earth's surface, but neither are the objects that are found on the earth surface to be studied in themselves-in geography-but rather the areas of the earth surface are to be studded in terms of the particular character resulting from the phenomena, interrelated to each other, and to the earth, which fill the areas."

In all of these respects there was no essential difference between Ritter's view of geography and that of Humboldt. Since Ritter had expressed many of these views as early as 1804 one may question Peschel's statement that it was in their first personal contact in 1807, that Ritter "was brought to see clearly the great function of geography to show the unison of human social phenomena with the complex of natural forces in the locality". Ritter may however have acquired his ideas from Humboldt's writings, with which he was already familiar, and the lengthy conversations with Humboldt over several weeks may well have emphasized and clarified his concepts. Furthermore Peschel is simply echoing many of Ritter's statements when he says that the latter's work in relating historical phenomena to physical phenomena would have been impossible had not physical geography been advanced so notably by Humboldt: "a Carl Ritter could not have preceded an Alexander v. Humboldt, but could only follow him".

On the other hand, the great difference in temperament and in general outlook on life gave a very different colour to the work of the two founders of modern geography, and in the works of their followers this difference became fundamental. It is natural enough therefore that later students should tend to emphasize the differences between the two men and even to find an opposition between them of which neither of them were conscious familiar though they were with each other's work. Interesting and dramatic as such a contrast may be, historical objectivity does not permit us to exaggerate differences that may be superficial into a fundamental contrast.

(Subsequent to writing this conclusion, I find that Ratzel had made much the same statement, emphasizing the many respects in which Humboldt's geographical work, even in its philosophy, was similar to that of Ritter.

In fairness to all varieties of critics of Ritter it must be said that in his voluminous and variegated writings one may find apparent justification for almost any possible Judgement. If at times he wrote as a scientist, at other times he wrote as a philosopher-indeed it did not occur either to Ritter or to Humboldt that science and philosophy could be completely separated—and at other times Ritter wrote as a theologian. Furthermore, Ritter was unfortunately far from careful in his selection of terms, "he did nothing to confine the fluid and ambiguous quality of words," "strict logic was not his forte". Any number of writers have debated what Ritter intended in calling his *Erdkunde, "allgemeine vergleichende"* (general comparative) and the net conclusion appears to be that we are not to look for any specific meaning in these terms, but merely to understand that he had certain general purposes in mind that he wished to contrast with those of his predecessors.

Ritter is therefore a writer who cannot be judged, either in his work or in his methodology, on the basis of a few striking quotations pulled out of their context and out of their relation to the ideas of his contemporaries. A view of Ritter constructed in this manner has the effectiveness, but only the degree of truth, of a caricature. We need not however make an exhaustive study of Ritter's work, but may base our appraisal on the many thorough studies of that kind made by previous students, in each case, however, referring their findings to the original work. The latter must include his actual geographic work as well as his methodological studies, since the methodology is expressed no less in the actual work itself.

In particular, one must avoid the error of attempting to determine the character of the geographic work of either Ritter or Humboldt—or of any other writer-from their own statements of ultimate purposes. The fact that Columbus set out across the Atlantic for the purpose of finding a shorter route to the Indies, and apparently died still believing that he had found it, does not cause us to judge his importance in history in terms of the establishment of connections with the Indies to that he contributed little more than a permanent confusion in geographic and ethnographic terminology.

Attempt to Discredit Ritter's Work : Writers of a later period, living in a different philosophical atmosphere, attempted to discredit the importance of Ritter's work by attacking his philosophical concepts, in particular his pietistic, teleological view of the universe. Commonly this is done under the naive assumption that the writer has a purely "scientific" concept, in contrast with a philosophical concept of the universe. We cannot here discuss this elementary philosophical problem, but merely point out that every scientist has his philosophy of science and that philosophy is neither science itself nor is it the product of science. That Ritter recognized this distinction may be seen particularly clearly in his discussion of Frobel's criticism of his work. To Frobel's objection to his teleological view which was not an argument but simply a different philosophical assumption of science asserted without foundation-Ritter replied properly in philosophical terms. Though we may reject his philosophy we can respect his statement of it. On the other hand, in reply to the attack on his scientific methods and concepts, he limits himself to scientific terms.

The proper question to ask in criticising the scientific works of any writer-as Plewe among others has noted is not: what particular philosophy of science is exposed by these writings, but rather: to what extent does the writer's particular philosophy affect his work. One may have a teleological philosophy-as did Leibnitz, for examples—and nevertheless still produce genuinely impressive research. Though many of Ritter's followers, like Guyot, depended on the teleological factor to explain geographic details, competent critics agree that Ritter seldom was guilty of this procedure which in fact he vigorously condemned.

In general, we may say that the teleology in Ritter's geography was an attempt to interpret philosophically that which science could not explain. Many of the statements at which the scientific critics of the late nineteenth century pointed in holy horror would stand comparison with current utterances of a number of eminent physical scientists who find themselves in a similar position. In Ritter's case, there were three fundamental facts of geography for which science had no explanation—and presumably still has none—namely, the uniqueness of the earth in the universe, so far as we know it; the earth as the home of that unique creature man; and finally—the fundamental explanation of a host of geographic fact the differentiation in character among the major land units of the world.

In our day, to be sure, it is more common for scientists to express their philosophical views of problems presented by their scientific work in separate publication; the intellectual world of a century ago did not call for this physical separation, nor can we regard it in itself as a major difference. "Even though a spirit turned toward the eternal breathes through Ritter's presentation," concludes a modern student, Schmidt, "even though his highest enthusiasm is fired by his religious *Weltanschauung,* nevertheless in his research, he in no way proceeds from preconceived opinions; his scientific procedure was directed throughout on temperate, purely factual comprehension of the facts and their relations. Ritter strove in the knowledge of the earth for a comprehension of the divine world plan in no other way than the natural scientists pursue the thought of evolution".

Although most critics would add some degree of qualification to Schmidt's Judgement, it seems fair to conclude that the disturbance caused in the modern scientist's mind by Ritter's teleological expressions is far out of proportion to their importance in his work. An illustration is offered in the case of his treatment of the continents, specifically in two lectures given before the Berlin Academy of Sciences, one in 1826, the other in 1850, republished later together. They constitute some 65 pages in which the author describes the size, form, construction, and climatic conditions of the different continents, seeks to construct "a law of the arrangement" of the parts of each continent to the whole, and particularly endeavours to show how these conditions have determined the development of the peoples of the different continents at different periods of history. One must look with a very careful and critical eye to find any statement that would disturb the scientific determinists of the late nineteenth century. Near the beginning of the first lecture, to be sure, Ritter expressed his belief that the earth had been planned as "the temporary nursery of the human species" and a sentence at the end of the second lecture suggests that a part of this plan was the particular form and position of each of the continents which had led as his detailed discussion had demonstrated to the particular function that each had played in the course of world history. Those who can overlook such expressions of belief will, like Humboldt, test the-essays in terms of their scientific content.

Humboldt's comment in this connection also throws light on his interest in human geography. Noting that he himself had long insisted on the great influence that the form of a continent exerts on its climate and vegetation, he discusses Strobe's effort to show the relation of the articulate and peninsular form of Europe to the development of its civilization, and adds: "In our time M. Ritter has developed with great wisdom the analogies, in physical and political characteristics, that the three peninsulas of Asia offer to the three peninsulas of Europe. The two groups offer centres of culture of very distinct physiognomy".

Nevertheless there were certain respects in which Ritter's philosophy did affect his work, and even more that of his followers. His purpose led always from the individual facts toward a "Whole" of all phenomena; rather than to trace the inner relations of phenomena to a "last cause," his purpose was to find the coherence of forces in a Whole, and thus ultimately to indicate the purpose of the Whole. This purpose must surely be found most particularly in the highest of earth creatures, the only one who could conceive of an organization of the Whole, namely man (it was logically false, he replied to Frobel, to suggest that it could just as well be found in cows. His consideration of the earth therefore logically centered on man.

This view, which may be found in his first geographical publication, in 1804, is expressed particularly in the statement of purpose with which he introduced his *Erdkunde:* "to present the generally most important geographic-physical conditions of the earth surface in their or possibly "its" natural coherent interrelation *(Naturzusammenhang)*, and that (the earth-surface) in terms of its most essential characters and main outlines, especially' as the fatherland of the peoples in its most manifold influence on humanity developing in body and mind".

While this concept of geography unquestionably places a major emphasis on the earth as the home of man, it does not restrict it to that particular view. "Independent of man, the earth is also without him, and before him, the scene of the natural phenomena; the law of its formation cannot proceed from man. In a science of the earth, the earth itself must be asked for its laws". The first volume of the *Erdkunde* gives detailed consideration to areas in Africa of apparently no concern to man. In the later volumes, however, Hettner finds that Ritter concentrated his attention on man to such a degree as to forget his purpose of establishing the total association of nature and man. Nevertheless one of his major accomplishments, according to Marthe, was to establish the physical foundations of geography in detailed study of what we would term regional landforms.

Further, Ritter's particular philosophical view-but also perhaps his failure to submit concepts to rigid logical examination-permitted him to further the traditional concept of regions as "natural divisions" of the earth—surface. That Ritter's natural philosophy has continued to affect, in this respect, the views of geographers who would deny that philosophy has been repeatedly suggested by Hettner.

Finally one may conclude that in his teaching, in contrast to his writing, Ritter's teleological views were more strongly impressed upon his students. There is some evidence for that in the lectures published shortly after his death, and far more in the performance of his students.

Unquestionably the modern scientist feels more at home in the writings of Humboldt than in those of Ritter. Peschel, though deducing from certain of Humboldt's statements that he could not have been a materialist, honours his avoidance of the confusion "not uncommon with the materialists as well as with their opponents, of presenting scientific investigations in some form of religious light". On one point, to be sure, Peschel finds that "the noble heart of Humboldt appears to have a bit corrupted his critical sense," namely in regard to race questions. Humboldt vigorously maintained the theory of the "unity of the human race," a concept that "came to prevail first through Christianity" and could be ignored on the slave-market only because of "the degeneration of Christianity by great wealth." The opposite "hypothesis of racial gradations among men" he characterized as "not only unkind *(lieblos)* but also false". Although Humboldt's thesis would find more support among anthropologists today than that on which Frobel as well as Peschel based their criticism, the criticism itself was essentially justified: his views on this question were strongly influenced by non-scientific considerations. In his youth, Humboldt had first found intellectual stimulus in what seemed to Goethe, Forster, and many others as the "barren environment of Berlin," almost exclusively in a small circle of Jewish intellectuals. From them he had learned of Lessing and Kant. He himself described the impressions made upon him by the degrading scenes of the slave-market, which he first witnessed in tropical America. Both in political circles and in his scientific writings he repeatedly and vigorously expressed his views on the race question. In his master work, the *Kosmos,* he chose to "close the general presentation of the natural phenomena of the universe" with the following quotation from his brother, Wilhelm v. Humboldt: "We wish to note one idea which is visible in ever increasing validity through the whole of history. . ., the idea of humanity, . . . to treat the whole of humanity, without consideration of religion, nationality, and colour, as One great closely related race, as one Whole existing for the attainment of one purpose, the free development of inner powers".

In 'common with most other students, Peschel fails to note the extent to which Humboldt's philosophical views find repeated expression in his scientific writings, nor does he note a certain similarity between those views and the concepts which Ritter expressed, commonly in a more religious tone. Geography was not, for Humboldt, a field studied as an end in itself, but rather as the means of comprehending "the harmonious unity of the cosmos" as a "living whole," "a unity in multiplicity". "Insight into the cosmic organism *(Weltorganismus)* creates a spiritual enjoyment and an inner freedom which even under fate's hard blows can not be destroyed by any external power.

For Humboldt, this unity of nature was no teleological, anthropocentric unity-he wrote of the comprehension of "the inner, *secret* play of natural forces", where Ritter spoke of discovering the "divine secrets"—but was a balanced unity of the whole of nature, of which man was a part. "In the forests of the Amazon as on the ridges of the High Andes, we perceived how" animated by one breath from pole to pole, only One Life is poured out in stones, plants, and animals, and in the swelling breast of man". An important factor in the evolution of Humboldt's thoughts was his close personal relationship with Goethe. With him, and with the entire generation of the "romanticists" (in the widest sense), Humboldt shared the idea of an organic coherence of all phenomena; this was a common characteristic of both of the founders of modern geography.

In other words both Humboldt and Ritter, in their philosophical point of view, were products of their time—in particular both were influenced by the thought of Kant and of Rousseau. To Ritter's religious nature the concept of a universe of order and law in which all phenomena of nature and man were interrelated required the assumption of some divine purpose and plan which the scientist should attempt to establish. Granted this assumption it appeared obvious that the ultimate purpose involved could not be simply the production of an intricate universal mechanism; the purpose could only be found in the life of the highest of the creatures of the universe, the only thing in the universe that was capable of recognizing either the order of the universe or a purpose behind it namely, man. In Humboldt, on the other hand, the same philosophical concept found a responsive chord in his feeling for the aesthetic rather than the religious. In the writings of Rousseau and St. Pierre, it was the descriptions of the "harmonies of nature" that appealed to him; through Goethe he received the concept of the *"Landschaft"* as conceived by the "romantic" movement.

Interesting as this comparison may be, between the philosophical viewpoints of the two founders of modern geography, it does not directly concern us here. The essential point is that both of them differed from their philosophical and literary predecessors in that both strove to demonstrate their philosophical concepts not by deductive logic, as in the case of Kant, nor by sentimental descriptions of subjective impressions of nature, as in the case of St. Pierre, but by objective descriptions of observations of nature. If there are exceptions in Ritter's work that we must overlook, the same is true, if in lesser extent, in the work of Humboldt. Throughout the writings of both students, the modern scientist finds repeated expressions of philosophical concepts that are now commonly excluded from such works. But modern science, which has had bitter fights with religious philosophy, has never come into serious conflict with the aesthetic view of nature; perhaps for this reason Humboldt's expressions of his *Weltan-schauung* are less disturbing to scientific readers today than are those of Ritter.

In general, neither Ritter nor Humboldt saw any conflict between science and philosophy. Ritter found that both geography and history "are directed toward the integration *(die Combination und das Mass)* of ideas and are therefore forced to philosophize". Likewise Humboldt, a student of the works of Kant and Fichte, felt a need for "something better and higher, to which everything could be referred" and therefore greeted with enthusiasm the natural philosophy which Schelling had founded. In a letter to Schelling, in 1805, and in publication two years later, he expressed his conviction that "a true natural philosophy could not harm empirical research"; on the contrary "such a philosophy leads findings back to principles and is likewise the foundation for new findings". Some years later, to be sure, when many students gave "pure thought" the preference over empirical studies and seemed to think that science could be produced by thought without research, Humboldt scathingly attacked this "mad" form of natural philosophy.

That Humboldt did not associate Ritter's work with the *"bal en masque"* of a philosophy that despised observations and experiments and taught "a physics that one can pursue without wetting one's hands" is indicated by a discussion immediately following his praise of Ritter's work in one of his lectures at the University of Berlin: "The description of the universe provides materials for a proper natural philosophy, the foundations for which are sought in many different ways. I can not find fault with these efforts, although I would be inclined myself to go to work more empirically. In this natural philosophy we need only fear and avoid with difficulty: false facts. Empiricists and philosophers should not mutually despise each other, for only bound together can we attain the highest goal".

Unity of Nature

For both Humboldt and Ritter, the concept of unity of nature presumed a causal interrelation of all the individual features in nature. The phenomena of nature were studied in order to establish this coherence and unity. For both it was axiomatic that the unity of "nature" included organic as well as inorganic, human as well as non-human, immaterial as well as material. The exclusion of any part would be not only arbitrary but would destroy the coherence and unity of the whole. In his first publication, in 1804, Ritter vigorously opposed the trend in "pure geography" that tended to overemphasize, he thought, the *"naturliche Landschaft,"* insisting that geography must describe and explain all the present conditions of an area. Ultimately, to be sure, his predominant interest in man tended to emphasize the human element to such an extent that he actually failed to accomplish his fundamental purpose.

Humboldt's view of science was limited only by the universe outside his own mind, *"Natur."* Goethe, as well as Ritter, called him "an academy in himself". Even if we eliminate those parts of his work that he did not regard as geography *(Erdbeschreibung)*, so extensive and rich which view of the field that almost every movement for some particular form of geography from the geophysics of Gerland to the aesthetic geography of Banse can find its antecedents in Humboldt.... None, however, may look to him as the precedent for excluding other parts of the field. "In the great enchantment of causes and effects, no material and no activity may be studied in isolation". "The highest goal of all observation of nature is the knowledge of our own nature: and therefore we conclude our description with a consideration of the races of men". Likewise the outline of systematic geography in the *Kosmos*—called "physical geography" according to the usage of the time—culminates in the section on man. The consideration of man includes spiritual aspects as well as material, "as though the spiritual were not also contained within the natural whole *(Naturganzen)*". Moral and aesthetic problems therefore formed a part of his considerations.

Regional Descriptions

It is true that in many of Humboldt's regional descriptions one will find little concerning man and his works, but he himself explained that his travels in equatorial America had taken him chiefly to areas where "man and his products disappear, so to speak, in the midst of a wild 'and gigantic nature". His purpose, in presenting "the general results that interest all enlightened men" is to describe not only the climate and "the appearance of the countryside *(paysage)*, varied according to the nature of the soil and its vegetable cover," but also the "influence of climate on organic life," "the direction of the mountains and rivers that separate the races of men as well as the plant societies" and "the modifications found in the condition of peoples placed in different latitudes and in circumstances more or less favourable to the development of their faculties. I do not fear that I have increased too much the number of objects so worthy of attention: for one of the fine characteristics that distinguish present civilization from that of past times is that it has enlarged the mass of our conceptions so that we realize more clearly the relationships between the physical and the intellectual world; hence there has developed a wider interest in objects that previously were of concern only to a small number of savants because they had been considered isolated from each other according to an overly narrow point of view."

In his descriptions of populated areas, Humboldt consistently strove to fulfil this purpose. Man, his culture and his works, are included as integral parts of his description and interpretation of nature. In his studies, of Central Asia, he was limited by specific restrictions imposed by the Russian government.

Both Humboldt and Ritter carried the concept of unity of all nature into the consideration of individual areas, at least in what I have termed the "vertical" sense—i.e., that all the features of an area in their interconnections form a naturally unified complex, whether or not that is to be considered as a unit Whole or merely a part of the one natural Whole of the world. For Humboldt, every part of the world is a reflection of the unity of the Whole. Ritter's statements on this point are more emphatic; both however, were, as we have seen, following paths already suggested by their predecessors.

How was this unity to be studied? It hardly seems necessary to say that neither of these "founders" of a science supposed that the study of an area was to be confined to a consideration of it as a whole, or that it should begin with such a consideration of the whole. While both assumed from the start that all the objects and forces of nature (nature in the sense of the world outside our minds) were interconnected to form a Whole, both recognized, in theory as well as in practice, that this could be established scientifically only by investigating the individual, single features in their relation to each other, and building these up in their actual relationships to form the whole. The function of the scientific geographer was to perceive these features, not separately, but in their interrelations so that he could thereby reproduce intellectually the unified whole that was nature.

That the previous statements correctly represent Ritter's purpose and method is supported not merely by the views of many other students more familiar with his work, but appears clear from even a brief consideration of either his methodological discussions or of his work. To be sure one may find sentences in his writings which appear to say the opposite. Thus a century ago the young student, Frobel, extracted one or two of Ritter's sentences from their context so that it appeared as though Ritter attempted to use synthesis without preliminary analysis—the "holistic" principle, as Leighly puts It, that "forbade the investigator to dissect a whole into its "parts for fear of destroying its coherence and so its essential character". That Ritter was not so naive as not to recognize that synthesis presumed analysis can read from the paragraphs that follow the quoted sentence on which Frobel has based his argument ; indeed every section of his work demonstrates the principle which he had definitely stated in his introduction, "that the procedure must lead from the' simple to the compound, from the individual parts to the unit". It was consistent with Ritter's personal character that he should have some hesitation in replying to the charge that his method "which builds up and combines" was one "in which analysis is unconsciously presumed". Many thought that he might have asked Frobel to read again the pages referred to, but he contented himself with the suggestion that it might have occurred to his critic that it would not be news to the author of a synthetic system that synthesis must be developed from analysis. Later critics, indeed, questioned whether Ritter had actually succeeded in getting beyond analysis to synthesis.

Horizontal Concept

The "horizontal" concept of unity of particular areas, as individual wholes, was also furthered by Ritter, though apparently with some qualifications. He did, to be sure, stress the individuality of the continents, which at times he referred to as "organs," but in his reply to Frobel he insists that the terms are not to be taken literally, but in the form of analogies (just as, one may note, Frobel had written of the systems of like phenomena on the earth as "organs". He was not entirely satisfied with the term "individual" and would drop it whenever he learned a better' one. Nevertheless he conceived of each continent as a "natural Whole"—in terms of all its characteristics. Similarly, he wrote, in some cases at least, of the divisions of the continents, *die Lander,* as "individuals" that are "members," rather than mere parts of the "organisms" of the continent. Although these concepts fit Ritter's teleological philosophy, they are not necessary consequences of that philosophy, nor is such a philosophy necessary to those concepts. In any case we are not to take the—terms literally.

Humboldt does not appear to have considered this concept. During concludes that "whereas for Ritter the individual area had its own particular determined value, for Humboldt it was only a variation of the great cosmic theme of law and causality". "Nature in every corner of the earth is a reflection of the Whole. The forms of the organism repeat themselves in ever different combinations". In a sense Humboldt was not faced with the question of regional division since he made no attempt to organize the regions of the world, or of any major part, into a system. That he recognized the need for such an organization of geographical study we will note later.

If we put aside the question of the philosophical viewpoints of the two founders of geography as irrelevant except as it affected their work-and in that respect by no means so important as one has been led to believe-what are the major differences between the directions which they gave to geographic work?

Although we find it erroneous to draw a fundamental contrast between them in regard to the proper content of geography, it is nevertheless true that the difference in character of their work led to an important difference in later interpretations. Though

Humboldt, at various places in his writings presents brief systematic considerations' in anthropological geography, his studies in systematic geography were primarily limited to non-human features. In Ritter, on the other hand, the interest in the study of non-human elements was subordinated to the interest in man, and his great work is presented in the form of a regional geography; further his methodological studies repeatedly emphasized the importance of regional organization of geography in contrast to that by classes of phenomena. In consequence of this double contrast we find the confusion between the two forms of dualism in geography to which was have previously referred-namely, that the distinction between human and natural (non-human) geography is contained in the distinction between regional and systematic geography.

If we separate these two issues, we find, as already noted, that the inclusion of human as well as non-human features in geography did not raise any question of dualism in the thought of either Humboldt or Ritter, but rather was absolutely essential to their fundamental concept of the unity of nature.

The second form of dualism, however, was unquestionably recognized by both men as a contrast in form of study and organization of geographical material. Furthermore, this dualism in methods has presented an issue that has recurred repeatedly in the history of geography, including the present time, and geographers have not been lacking who wished to solve it by the simple method of excluding either one or the other aspect-that is, by limiting geography either completely to systematic studies or completely to regional studies What was the position of the two founders of modern geography with reference to this issue-as expressed both by their theoretical statements and in their practice?

There can be no question that Ritter's emphasis on regional organization, both in his work and in his methodological statements, tends to give the impression that he had little interest in systematic studies of each analyzed element over the world. But if our comprehension of a student, generally regarded as of major importance in the development of geography, is to consist of more than mere impressions based on the outline of his work and a few sentences extracted from his methodology, we must examine this question more carefully than did the young critic of a century ago.

The present issue is not to be confused ;with the related question, previously discussed-the argument that Ritter attempted to synthesize the factors associated in an area without preliminary analysis. That question might be dismissed with the mere statement that no matter what he may have appeared to say, in a few sentences removed from their context, the man whom successive generations of eminent scholars regard as a master of geography was no fool. The present question, however, is a genuine one.

Ritter's methodological discussions are 'to be regarded less as balanced treatments of a field of science than as the program of a reformer attempting to reconstruct a previous geography into a science, In the systematic studies of certain of his predecessors it seemed to him that the intellectual ("subjective") process of dividing the interrelated phenomena of nature into separate classes and studying these separately over the world was a disruption of the actual coherence of nature—a conclusion, we may note, that requires no teleological basis. He observed that, when each of these kinds of phenomena is studied in its forms and processes, the result is not only comparable with that of other sciences, but is rather a part of some other science. A "compendium" of these separate studies would not form a geography and the aggregation of areal sections of each of them for any particular area would be no science at all.

Ritter therefore insisted repeatedly that the proper method for a science of the earth was "to ask the earth itself for its laws" ; it must use the "objective" (we would say "empirical") method, "calling attention to the main types of formations of nature"; by investigating the relationship grounded in nature itself, this method leads to a "natural system".

On the basis of such statements, then, one might, like Frobel, accuse Ritter Of ignoring the importance of systematic studies of ,individual features. To be sure, one finds statements which indicate the opposite, but one must look fairly carefully to find these. For example: "the earth's surface, its depths and its heights, must be measured, its forms arranged according to their important characters". Further in the lectures given before the Academy of Sciences in Berlin, in 1828 and 1833, one will find considerable discussions of the need for studies of if individual categories of phenomena.

Nevertheless, most later critics have felt that Ritter's regional work would have been more effective had it been based more thoroughly on systematic studies, even though the resultant delay would have limited its scope. This criticism was never effectively presented during Ritter's lifetime. It might be inferred in Bucher's remarkable critique of 1827—to which we have already referred-though that is certainly not directed at Ritter. Whereas the *Erdkunde* is barely mentioned, Ritter's earlier systematic geography of Europe is treated at length as the principal example of the correct systematic approach. But whether Bucher intended to imply a criticism of Ritter's later work, or whether Ritter was conscious of that, I do not know; the work was certainly called to his attention by an editorial footnote to his own article in *Berghaus Annalell,* in 1831. Neither Marthe nor Wisotzki, both of whom recognized the importance of Bucher's study, appear to have inferred any criticism of the *Erdkunde.*

Frobel's Essay : Much better known is the critical essay by Julius Frobel, at that time a young student with two years' experience 10 academic geography. On Kitten recommendation, Berghaus published the article in his *Annalen* and the fact that Ritter commented on it at length in the same issue focussed the attention of the academic world upon it. Describing the episode some fifty years later, Frobel speaks of his surprise and embarrassment at finding himself, a young student of no particular attainments at the time, thrust suddenly into the academic limelight as an opponent, of the master. It is an interesting indication of Ritter's character that in his published marks about Frobel's critique, he ignored the easy opportunity to make Frobel look foolish, and on the contrary commended his critical efforts as an expression of an - "earnest thought for the furtherance" of geography from an "unmistakably keen mind". Further, when Frobel came to Berlir shortly after, Ritter received him cordially and the young student received much needed employment working on the maps for the *Erdkunde.* Frobel's account of his reception by Humboldt is also interesting, but his statement of the latter's apparently favourable reaction to what he himself termed his "impertinent" criticism (of Ritter) is not specific and has only the degree of credibility granted to personal reminiscences written more than fifty years after the event.

In spite of the attention which Frobel aroused, the net effect on the geographic thought of either Ritter or his contemporaries, appears to have been nil", Whatever degree of truth there may have been in Frobel's criticism was lost sight of because his attack, based on but limited knowledge of Ritter's work, so widely overshot its mark-in striking contrast, we may note, to Bucher's critique which had been based on very careful analysis of the works criticized; Frobel does not mention Bucher's study, not even in his succeeding article, after Berghaus had rather pointedly called attention to it.

Although it was Ritter's "point of view" and "method of treatment" in geography, that Frobel subjected to criticism, his consideration ignored Ritter's methodological essays and confined itself to the two volumes of the *Erdkunde* that had been published up to that time. On the basis of a few sentences in the introduction of the first volume some of which he misunderstood, Frobel arrived at a critical conclusion concerning the value of the work as a whole. In one case, as previously noted, the criticism is valid only if one considers the single sentences removed from their context. Frobel's discussion of Ritter's use of the term "comparative," together with the latter's comments in reply, did little to clarify that question, since, as Plewe notes, each party proceeded from different assumptions and ignored those of the other. Undoubtedly Ritter's lack of precision in choice of words contributed further confusion to a problem that had previously bothered geographers, and over which many later students have wrestled; the most thorough and illuminating discussion of the whole question is presented in Plewe's *"Untersuchungen uber den Begrillder 'vergleichenden' Erdkunde"*.

On the issue with which we are here concerned—the relation of, systematic studies to regional geography-Ritter's theoretical statements offered clear grounds for argument. In reaction to those who had specialized on systematic studies and never reached the actual formations of interrelated factors in areas, he had called first for the study of the areas filled with interrelated phenomena, asserting that, when this had been done for the whole world, the material would then be available for a more successful development of the general principles of systematic geography. Ritter maintained his belief in this methodological principle in many later writings, notably in his essay of 1833.

In assuming, however, that these theoretical principles constituted the foundation for Ritter's *Erdkunde,* his critic ignored, or was ignorant of the fact that elsewhere in his methodological discussions Ritter had shown the need for the opposite method, and that in practice he had in large part followed it. We have already noted his early study of the major geographic features of Europe that Bucher had cited as the model for systematic studies, and the handbook of general (systematic) geography. Although the latter had not been published, various sections had been used in different school texts, in some cases with the original authorship indicated, Ritter could therefore point to these, in his observations on Frobel's critique, as well as to the systematic discussion of individual categories of features that fills more than fifty pages of the introduction to the *Erdkunde.* Whereas Frobel had overlooked these, Humboldt and other writers had found in them general concepts and principles that could be used in their own geographic writings-as Ritter observed with detailed references. In subsequent volumes, we may add, Marthe has counted no less than twenty-four systematic studies, some running over a hundred pages each, dealing particularly with minerals and cultivated plants; in many other cases, however, Ritter refers to studies to appear in later volumes that never 'appeared.

Further, the essays read before the Academy of Sciences in 1826 and 1828, include methodological discussions of systematic studies. In the latter, for example, there is a discussion of the study of geometric forms of areas of, different categories of phenomena, an early example, we may note, of studies of "patterns," Finally, Ritter could point to his lectures in general (systematic) geography that had been heard by hundreds of his students during the preceding decade. Although he had not published these, they had actually appeared in modified form as Berghaus' *Elementen der Erdbeschreibung*—the very work that Frobel had praised as the model of what should be done in geography. Berghaus himself, as editor of the *Annalen,* confirmed this statement in two emphatic footnotes-not without implying that the fact should have been sufficiently clear to any reader of the *Elementen.* The irony goes even deeper than Ritter recognized. In repeating the statement more clearly in the introduction to the first volume on Asia, the following year, Ritter claimed that the inner organization of Berghaus' work was the result of his lectures, but gave his friend credit for having improved the style of presentation for elementary students. Ludde however found, by comparing the originals, that Berghaus had published straight dictation of Ritter's course—on many pages, word for word.

It is interesting to note that Ritter, who in his youth had been scathingly ridiculed by one reviewer for presuming to criticize his elders-in a work that at the same time offered positive materials-should have limited his personal criticism of Frobel to the polite but pointed request that he should make himself more familiar with a man's work before attempting to subject it to a critical Judgement. It is not surprising, therefore, that Ritter's contemporaries should largely have ignored the degree of truth included in Frobel immature and misinformed critique. On the contrary many appear to have followed the more zealously the methodology that Frobel had attempted to criticize. Undoubtedly this was impressed upon them by the constant emphasis in Ritter's methodological treatments on the full studies of areas, as well as by the contrast between the nineteen volumes of the *Erdkunde* and the small amount of distinctly systematic studies published under his name. With the notable exception Redur they did little to further the systematic studies that Ritter had made, and in their regional studies they depended even less than he on systematic materials already available. The fact that so much of the work of Ritter's followers later came to be regarded as of little value might be taken as a demonstration of the argument that Frobel had directed at the *Erdkunde:* that it could not attain scientific value because it had not been preceded by systematic studies. Indeed, the argument against the *Erdkunde,* itself, though evidently exaggerated, was perhaps ultimately vindicated. Although scientific travellers, as well 1'-8 geographers, for many decades-in Germany and especially in Russia-found in Ritter's work "the essential register of facts on Asia", yet for modern students it appears to offer little of importance even to the historical geographer (as distinct from the historian of geography). On the other hand, one may remember, Ritter had found that his earlier attempt to provide first a systematic geography of the world had not been regarded by Buch as successful. Consequently the relative success or failure of Ritter's work does not lead to a clear conclusion as to the proper order of study of systematic and regional geography.

View on Humboldt's Approach : Humboldt's approach was undoubtedly different in many respects. Even less than Ritter, did he regard geography as an end in itself. In order to establish the unity of the total cosmos, it seemed more important at least for the time-to make systematic studies of particular kinds of phenomena in their interrelation in areas, than to prepare complete studies of individual areas. Though he made many of both types of studies, his effort to cover the whole of "physical geography" in the *Kosmos* demonstrated his earlier statement that he regarded this as the more important of his two purposes. In referring to that well-known quotation however, one must understand clearly what he meant by the term *"physisehe Erdbeschreibung"-the* expression which he finally adopted after trying various others, such as "physics of the world." We have already noted that in its scope it included all earth phenomena outside the individual student's mind: man and his culture, as well as other organic and inorganic phenomena. Further, it was not the individual categories of phenomena in themselves that primarily interested him, though he had been "passionately devoted to botany and parts of zoology," but rather the interconnections of different categories of phenomena—i.e., of plants of different kinds with each other, and with the differences in climate, relief, and soil and in relation to animals and human life. In addition to the types into which everything that *is,* may be classified in terms of its intrinsic characteristics, one perceives types in the arrangement, distribution and mutual relations of things of different categories. "It is the great problem of physical geography *(physique de monde)* to determine the form of the types, the laws of these relationships, the eternal bonds that enchain the' phenomena of life with those of inanimate nature". Throughout all his geographic work Humboldt remained true to the principle he had set for himself in a letter written on the day of his departure for America: "On the interrelation *(Zusammenhang)* of forces, the influence of inanimate creation on animate animal and plant world, on this harmony, my attention shall always be directed" quoted by Richthofen. More specifically, in the first volume of the *Kosmos,* Humboldt described physical geography as the study of "that which exists already *(im Raume)* together it studies phenomena as arranged in areas in their mutual relations to all other phenomena with which they form a natural whole.

In contrast with those who attempt to construct geography after the pattern of other sciences, Humboldt regarded it as having a fundamentally different viewpoint from all the other sciences concerned in the study of nature. The individual natural sciences what we would call the physical and biological sciences concerned with earth phenomena-study the forms, construction and processes of individual animals, plants, solid objects or fossils and seek to arrange these in classes and families "according to their internal analogies." Geography is concerned with these objects as they exist together, related to each other causally in an area *(Raum).* In contrast with physics or botany, in which "the objects of nature are divided according to kinds of objects," geography regards all the objects as a natural whole, as they stand in areal connection, in part with the earth body, in part with the universe. Cosmography *(Weltbeschreibung-the* term is taken from Kant) is "the study of all that is (the things and forces of reality) in the area, as a simultaneously existing *Naturganzen."* It includes both an astronomical part and a geographical part. Finally, geography is closely related to the third point of view in studying the earth—more closely than to that of the specialized science—namely, the historical, which is concerned with the changes in time of all the phenomena.

Similar statements of the comparison of geography and history may be found in Ritter's writings, including his first publication of 1804. In particular he notes that as chronology provides the framework into which the multiplicity of historical facts are ordered, the area *(Raum)* provides the skeleton for geography; both fields are concerned with integrating different kinds of phenomena together, each in its respective frame. To what extent he was indebted to Humboldt for these ideas we cannot tell; neither is it clear to what extent each may have depended on Kant, who had long before expressed similar ideas in the lectures on physical geography that were not published until 1801-2. (Humboldt refers to Kant in his lectures of 1827-28, but his own first statement of the concept of geography, on which later versions are based, was published in 1793.

Heterogeneity Feature : Heterogeneity therefore is an inescapable characteristic of geography, as Humboldt understood it. On the other hand, as an eminently practical scientist (by which we do not mean an applied scientist) he recognized that to comprehend a whole consisting of a multitude of things in interrelation it was necessary to understand first the relations of some of the things to all the others. His studies in systematic geography, therefore, concentrated on particular classes of phenomena-particularly vegetation—in relation to all other phenomena in areas. Whereas "botany, proper studies the character, organic formation and relations of plants, In the geography of plants' descriptive botany is tied to climatology." The development of plant geography required the combination of means of measuring elevation and temperature with a knowledge of plants, and had been delayed until the methods of the physicist could be combined with those of the botanist. In plant and animal geography however "we speak not of the plants and animals" but of the earth-crust covered with plants and animals".

In other words, Humboldt recognized that, though a systematic study in ,geography must be focussed on one category of objects, it was in no sense to be restricted to the objects of that category but was to consider those objects in their relationship to other geographic phenomena. His physical geography therefore cannot be completely divided into a list of highly specialized fields each separate from the other; rather, in each part, one group of features is the centre of attention, but its relations with others carries the work into all the other divisions. Since these relationships are those based on areal position-rather than historical sequence—the fundamental unifying principle is chorological.

This chorological point of view Humboldt consistently maintained in all studies that he entitled geography. Even in a textbook for geography that he and Berghaus projected (for schools in India, at the request of certain English authorities) he refused to insert material that properly belonged in botany or zoology. "A physical geography can concern itself neither with energy and material, nor with the physiology of the organic bodies; all that must be assumed as known".

Analysis of Humboldt's Contribution : Humboldt's contribution to the development of systematic geography is therefore to be found not only in the value of the works which he produced but also in the fact that he first clearly portrayed the' distinction between systematic, but chorological, studies in geography and systematic studies in the special sciences. That he was for a time regarded as an important figure in such sciences as botany or geology rested on a misunderstanding of his work, as he himself explained it. The Judgement of later specialists in these fields, that Humboldt had done little of major importance in their sciences, was an echo of his own purpose. The fact that, rather than attempting systematic studies in the special fields; Humboldt, through the greater part of his work, maintained the chorological point of view, enabled him to found several branches of systematic geography, notably climatology and plant geography.

On the other hand it would represent a complete misunderstanding of Humboldt's concept of the field of geography to suppose that he thought it limited to what he called "physical geography," *i.e.*, to chorological studies in systematic geography. Since his own work was based on his field observations-enriched by those of a great host of other writers, both scientists and travellers-he never prepared a regional geography of the world or even of a single continent. But his recognition of the need for such studies, coordinate with work in systematic geography, is shown by the outline which he drew up in 1848 for the geography text to be used in India, in which" *Specielle Geographie: Landerbeschreibung"* forms the second half, coordinate with the first, *Physikalische Erdbeschreibung*. The same view is indicated in his discussion of Varenius' outline of geography as well as in his many references to the work of his "great and celebrated friend," Ritter. More significant, however, is the large number of explanatory descriptions of individual regions which he studied in terms of the interrelated totality of their phenomena.

Many of these regional studies of Humboldt are interspersed in the record of his travels others will be found in his *Ansichten der Natur* an excellent collection of such regional descriptions has recently been published in a single volume of selections from many different writings by Humboldt. If Humboldt considered these as less important than his systematic studies, he certainly did' not regard them as unimportant,' as is shown by his careful and detailed discussions of the difficulties involved in such studies and of his method of solving them.

For Humboldt, "there is a certain natural physiognomy that belongs exclusively to each climatic zone; every vegetation zone, in addition to its particular advantages, has' also its particular (or specific) character." His purpose was to gain "a comprehension of the natural character of the different regions of the world "which "is most intimately related with the history of the human race, and with that of its culture". Although Humboldt's personal inclinations may have led him to prefer to study areas in which man was of little importance, his interest was limited neither to the nonhuman features nor to the visible 'landscape features. On the peninsula of Araya (now in Venezuela) his range of observations included astronomical and atmospheric conditions, plants and animals, the salt marsh and the salt works.-based upon it, the pearl industry, the physical characteristics of the various races represented, and particularly the changes in the character and customs of the native-born Spanish colonists resulting from the differences in natural conditions as compared with Spain. This situation he contrasted with that found in the highlands, or in the United States, where European settlers had found climatic conditions more similar to those in Europe, and with the character of the ancient Phoenician and Greek colonie.

Similar considerations may be found in what might be regarded as Humboldt's more scientific works. "How powerfully has the sky of Greece influenced its inhabitants! Were not the developed who settled in this beautiful and fortunate part of the earth, between the Oxus, the Tigris, and the Aegean Sea, first awakened to moral grace and sensitive feelings? And did not. . . our ancestors bring anew gentle manners from those gentle valleys ?" "The influence of the physical world of the moral, the mysterious interrelation of the material and the immaterial, gives to the study of nature, when one raises it to a higher point of view, a too-seldom recognized charm of its own". Peschel appears to have overlooked such statements in Humboldt's writings while objecting to similar suggestions in the work of Ritter's followers, specifically of Kapp.' In particular, Peschel attacks the assumption of "a causal relation between the impression of the *Landschaft* and the mental expressions of the population"

On the other hand, Humboldt's descriptions ,of areas, Peschel found, had introduced a new period in regional geography *(Landerbeschreibung).* He succeeded not merely in presenting the picture of the area, but also in "making it live through the play of forces;" thus he "made real the interconnection of the observations with a higher order of the Whole." "The post-Humboldtian geographer is concerned not only with the interrelations of physical phenomena, but also with those of the historical human] phenomena with their scene." In particular, Peschel finds hat geography was raised to scientific stature in the regional studies of Cuba and Mexico. In these studies, to be sure, Humboldt's interest in natural conditions is subordinate to that in social conditions and much of the material, according to Hettner, is not to be included in geography. Humboldt himself apparently regarded these as studies in political economy. Nevertheless the manner in which he described the differences in economic, social, and political conditions of different areas in relation to the differences in natural conditions was in fact followed, not by economists nor political scientists, but by geographers, who thereby, according to Peschel, made this form of regional study one of the functions of geography.

In his study of the development of economics and of economic geography, Schmidt entitles Humboldt's study of New Spain "the first scientific, political and economic geography founded on the nature of the land". Paradoxically we may note while Humboldt thus made a major contribution to regional economic geography, Ritter, in his incompleted studies for a *"geographische Produktenkunde,"* was one of the first to make systematic studies in economic geography.

Differences between Two Founders : In general, we may recognize two major differences between the two founders of geography in their actual development of regional geography. Ritter's ambition to produce a regional geography of the world—even though his long life did not permit him to carry it beyond Africa and Asia-led him to consider regions of relatively large size in which he seldom if ever completed a full study of all related Phenomena. In contrast, Humboldt described in great detail areas he had visited, some quite small, and so was able to provide-within the means of observation then available-a relatively complete picture. In order to accomplish this he not only considered each of the significant features of an area but combined these individual characteristics for each distinctive part of the area, then showing the relation of the different parts of the area. Indeed the great attention which he gave to certain small 'areas would justify the students of "micro-geography" in proclaiming Humboldt as their leader.

The second, more important, difference we have already indicated. Ritter maintained in theory, and in part in practice, that geography' should' first study all the interrelated phenomena to be found in each of all the areas of the world, on the basis of which systematic studies could be made of the relations of individual types of phenomena. The opposite view, that Frobel proclaimed, called first for a complete study of the individual phenomena systematically over the whole world. Humboldt, as a practical scientist, evidently felt he must do what he could do: he could more easily consider the relations of the two sets of phenomena—*e.g.*, plants and elevation-than those of all phenomena found together; furthermore he could not wait for a complete world regional survey. On the other hand he felt called upon to provide an explanatory description of the lands which he had studied as fully as he could study them. For this purpose he could not wait until systematic geography had been completed. But, by using the results of his systematic studies in his descriptions of areas, he produced "masterpieces" of regional interpretation.

Result of Differences

The ultimate results of these differences are both interesting and paradoxical. Ritter's relative neglect of systematic studies not only detracted from the value of his regional work, but set a tradition which for a time limited geography in such a manner that regional geography itself could not progress. In the case of Humboldt, on the other hand, the great thesis of the cosmos, that was of first importance to him, is now of little moment' either in science or philosophy. Though his studies in systematic geography were of major importance in the historical development of the field, they have long since become obsolete. In part the same is true of his studies in what has come to be called "comparative regional geography" in which he, rather than Ritter, was the pioneer. But that part of his work which he once indicated as of lesser importance, the masterly explanatory descriptions, analytic and synthetic in form, of individual regions of tropical America and Mexico, remain of imperishable value in geography. As Lehmann noted of any successful studies in regional geography: they cannot become obsolete, they provide irreplacable material for historical geography.

Brief consideration may be given to certain other effects of the work of Humboldt and Ritter on the subsequent direction of geographic thought. Humboldt was always greatly interested in the aesthetic aspects of geography, "as a means of stimulating and widening scientific nature study" Though he found models of such description in the writings of Goethe, Rousseau, de St. Pierre, and various French and English travellers and literary writers, previously mentioned, it was particularly his "illustrious teacher and friend, George Forster," who had "most effectively pioneered in this direction. Humboldt insisted that this aspect of geography was not to be distinguished from scientific geography; he was evidently unmoved by Frobel's specific demand that such studies should be grouped with Ritter's historical-geographical studies in a non-scientific "historical-philosophical geography". Although he does not mention Frobel's essay, his statements in the second volume of the *Kosmos* present a specific reply to the former's suggestions. It was not the description of his own feelings that the author of nature descriptions was to present; on the contrary by describing objectively the external nature surrounding him he should leave complete freedom for the feelings of the reader. The geographer's descriptions of nature, we may conclude, are not to be artistic in the sense of expressionistic; neither are they to be artistic in the sense of form. The aesthetic aspect is not to be supplied by "poetic" phrases of the writer, but by the actual scene itself which it is merely his function to reproduce as well as his command of language may permit. "Descriptions of nature can be sharply limited and scientifically exact without thereby losing the living breath of the power of imagination. The aesthetic aspect must proceed from the presentiment of the interrelation of the sensual with the intellectual, from the feeling of the universality, reciprocal limitation, and unity of nature-life." It is not alone the description of the view naively beheld that is to provide the intellectual pleasure, but far more the understanding of the "harmonious interplay in the forces in the landscape". Dove, however, felt that Humboldt had himself been guilty of employing the poetic form of description to which he objected in theory.

Critical Analysis : Neither Humboldt nor Ritter give a categorical answer to the question of the physical scope of geography-whether it includes the earth as a whole or is limited to the earth surface. No doubt one may find in Ritter's writings, as Gerland was able to do, statements that indicate the former, but it is clear that these are in conflict with the greater part of his methodological discussions as weli as with all of his work: the "earth" which he described as the show place of the forces of nature, and particularly as the dwelling place of man, was not the earth body but its enveloping shell. It was characteristic of Ritter that he should have been primarily responsible for establishing the term, *Erdkunde* or earth science—a term, as Hettner observes, that so poorly expressed his own concept. Whereas for Ritter the word was used as an alternative to the word *Geographie,* of foreign origin, and to the less scientifically sounding *Erdbeschreibung* which Humboldt favoured, later students came to interpret it literally as the science of the earth.

Humboldt, in order to establish his thesis of the unity of the cosmos, took the entire universe in his view, but within this distinguished "*Erdbeschreibung*" as the "telluric or earthly *(irdisch)*" part. His own research was largely limited to the earth, and indeed, to that part of it which he knew-the earth surface. Consequently the tendency of the pre-classical geographers to confine their field to the earth surface was, in practice, continued through the classical period.

In one other respect, Humboldt's actual work extended beyond the field that Ritter considered, namely in the study of the distribution of different types of phenomena—essentially one may say, "the Where of things". He may, however, have regarded such distributional studies as merely preliminary investigations necessary for the systematic studies of interrelated phenomena; in any case they constitute but a part of his geography.

Finally, we may add, in the development of the methodology of geography during the middle of the nineteenth century if not longer-two particular factors caused Ritter's influence to be of much greater importance. As the holder of the only university chair of geography in Germany-if not in the world-he exerted a great influence beyond the range of his immediate students, whereas Humboldt's influence was for some time greatest with men who did not consider themselves geographers. Secondly, whereas Ritter repeatedly expounded his views on the nature and problems of geography in methodological papers, Humboldt's numerous discussions of such questions were scattered through his general writings. For this reason, perhaps, his concept of the relation of geography to other sciences was lost sight of for nearly a century.

Academic Geography

The "classical period" in the development of geography may conveniently be considered as terminating with the death of both Humboldt and Ritter in 1859. During the following decade academic geography was dominated by the school which Ritter had founded. Since his followers tended to emphasize the "historical" aspects of the field even more than had Ritter, one may speak of a drift away from systematic geography to a regional geography primarily concerned with man. According to Peschel, Ritter's "single laudable student" was Ernst Kapp, but Kapp's interest extended also into political problems, and following 1848, disciplinary action led him, like Frobel, to come to the United States, where he settled as a cotton-farmer in a German community in Texas and apparently did not again engage in geographic studies until after his return to Germany in 1865. Another follower, if not student, of Ritter to come to the United States, were the :French Swiss, Arnold Guyot, who held perhaps the first chair of geography in this country, at Princeton. That he had but little influence may perhaps be accounted for by the manner in which the teleological view dominated even his interpretation of details.

By far the most successful of Ritter's students and followers was the French geographer. Elisee Redus. According to Girardin and Brunhes, it was his study under Ritter, in 1851, that made Reclus a geographer; from Ritter he derived his main principles and ideas concerning geography. It is interesting to note that Reclus also followed Ritter's practice, rather than his theory, in first preparing limited studies in systematic geography and then a general systematic geography, before proceeding to a complete regional survey of the world. Unlike Ritter, however, he established his reputation by his systematic study, a major work in physical geography *(La Terre,* 1866-67; later translated into English). Although Girardin and Brunhes found that the influence of Ritter especially predominated in this work, other students have noted the dependence on many other writers, including the English student, Mary Somerville. Sporer praised this work as superior to any of its kind in the German literature, contrasting it with the work of the Ritterian school, apparently not recognizing Reclus' as a student and follower of Ritter; presumably the same is true with regard to Peschel's statement quoted above. Hettner however finds the work distinctly limited by Ritter's influence'. Schmidt's statement that Reclus "became the Ritter of France" no doubt is based largely on his nineteen-volume regional survey, *Nouvelle geographie universelle.* 1875–94. Though this obviously followed Ritter's *Erdkunde,* it was far more successful, "both in the taut organization, which permitted him to complete the great work, and in the close coherence of nature and culture in every earth area".

Ritter's Influence : Ritter's influence was also significant in certain other fields. One of his most interested students at the military college was Moltke, who published various geographic studies long before his military-geographic plans became of momentous importance. Likewise Ritter was influential in calling the attention of historians to the significant relation of geography to the course of human events; Marthe notes this influence particularly in the work of the celebrated historian E. Curtius, a former student of Ritter's.

Though Humboldt, who held no university post, had no immediate followers in academic ranks, his influence outside the universities-and outside of Germany—was vastly greater than that of Ritter. During the brief "post-Ritterian period" Hettner finds that "the real representatives of the true geographical science were the scientific travellers who took Humboldt for their model".

Post-Ritterian Period : The post-Ritterian period however proved to be but a brief interlude before the last quarter or third of the century brought a very rapid development in academic geography in Germany. In many respects this period may be regarded as the critical period in the development of the field. The foundations which Humboldt and Ritter had established for geography did not provide, in appearance certainly, a clearly unified field. To the extent to which their followers exaggerated certain aspects of the views of each of the founders, or attempted to introduce new concepts of the nature of the field, geography was for a time split in several directions and its position as a branch of knowledge thereby brought into serious question. Following the death of Ritter there was no professor of geography in any German university and the return to university status and particularly the rapid subsequent growth was largely the work not of the "historical geographers" who followed Ritter, but of students who had been trained as geologists and tended to specialize in the study of nonhuman features of the earth—*i.e.*, physical geography as we understand the term. With the rise in academic status of geography and the productive work or this period we are not here concerned; in the development of geographic thought, its major problem was to overcome the apparent disunity in the methodology of the field and thus definitely to establish its position as a single field of science.

The general scientific atmosphere during the latter part of the nineteenth century was far from receptive to the philosophical concepts of the earlier "romantic" period, whether those of Ritter or of Humboldt. This change was evident even before the end of the classical period. Indeed the differences in geographic thought overlap our arbitrary divisions of its historical development to such an extent that we find certain of the concepts of the postclassical period expressed by the last of the pre-classical geographers, Bucher. We have already noted 'Bucher's thorough attack on the use of "natural boundaries" and "natural regions" in the works of his contemporaries. Utilizing anatomy and physiology as analogies-though recognizing that the analogies were not complete-he arrived at the negative conclusion that geographers need not attempt in any way to divide the earth into areal parts except for special purposes; rather that they should study it in terms of classified phenomena, *i.e.*, systematic geography. Although Bucher used Ritter's study-of 1806 as an example of what should be done, and speaks in other connections in praise of Ritter's work, his criticism unquestionably was applicable to the volumes of the *Erdkunde,* whether he realized that or not.

Though Berghaus appears to have been influenced by Bucher's study, and Marthe was evidently familiar with it, it is not clear that it had any important significance in the development of geographic thought, possibly because of its publication by a small provincial press. On the other hand the very similar views which Frobel presented a few years later in the principal geographic journal of the time created quite a stir in academic quarters, particularly because they were directed at Ritter and were accompanied by Ritter's comments.

The controversy however proved to be flash in the pan; Frobel's demands "died away without effect" and Hettner therefore omits them from his consideration of the development of geographic thought. That view may be somewhat exaggerated however, since. Frobel's articles, at least the two published in Berghaus' Annals, retained a fairly conspicuous place in geographic literature and may well have influenced later students whether they realized it or not. Certainly the viewpoint which he was trying to express ultimately came to the fore and Plewe therefore regards him as a herald of the new period in geographic thought.

New Viewpoint

The new viewpoint in science which came to dominate scientific thought, Plewe characterizes as marked by the increasing specialization of the sciences, by an increasing emphasis on the development of "scientific laws," and by a conscious isolation of science-and specifically of geography—from any particular *Weltanschauung:* this latter point of view, however, often represented an equally definite philosophical presumption of a materialistic., mechanistic, universe in which man was to be studied as a "thing," like any other—a sum of atomic movements".

The shift in geographic work is generally regarded by German geographers to have been due primarily to the work of Peschel and Richthofen. The movement was originated in the essays which Peschel published from 1866 on, a number of which were collected in 1870, in *"Neue Probleme der vergleichenden Erdkullde als Versuch einer Morphologie der Erdoberflache.* "It has been quite justly said" writes the Belgian geographer Michotte, "that with this work the scientific spirit reentered geography". Through this work, and through his teaching, Peschel led geographers to study primarily the morphology of landforms. His geographical work was by no means limited to this part of the field. On the contrary he was concerned also with the study of the influence of landforms on human history, and was successful, Marthe found, in limiting this more narrowly and sharply than Ritter had done. His early death, at the age of forty-nine, cut his work short just at the time when his university position should have enabled him to be most productive. (For a more adequate treatment of Peschel's contribution-as wen as a biographical sketch-see the study by Ratzel, as wen as briefer statements by Schmidt, and Doring).

The scientific morphology for which Peschel had striven was given its foundations by geographers who had been trained as geologist—notably by Richthofen, in his studies of China, published from 1877 on. Among the followers of Richthofen in this particular direction we may mention two: Penck-also trained as a geologist-whose classical study of *"Die Morphologie der Erdoberflache"* first appeared in 1894, and whose many students are still active in this field; and the American, W. M. Davis part of whose work, as we have noted, is included in the German literature.

In consequence there was inaugurated in Germany as later under Davis in America a long period when geomorphology represented the major field of geography. The inclusion of this subject within geography appears thereby to have been irrevocably established in Germany, whatever logical objections might be raised, whereas in America, its position with reference to geography and geology is still uncertain.

Whatever views might be held in regard to geomorphology, it has been commonly recognized that under Peschel's leadership, geography for a time tended to expand in the natural sciences into fields that had long been claimed and cultivated by other sciences, so that it seemed for a time as though geography would claim all of physical science that was related to the earth. Human geography on the other hand-including ethnography, agricultural land use, trade, and the movements of peoples, as in Richthofen's studies of California' and of China was considered largely in relation to landforms, or was confined to studies in regional geography, In the development of regional studies particular mention may be made of Kirchhoff, as a connecting link from Ritter to Ratze.

One important result of this situation was the confusion, to which we have previously referred, of two forms of dualism in geography into one: a systematic, physical (now meaning nonhuman) geography, and a regional, human geography. In order to justify the inclusion of studies of man in geography, and at the same time to prevent the indefinite expansion of geography in that direction it seemed necessary to state the purpose of the two parts separately. Natural features were studied in their own right-presumably it mattered little how far the field expanded in that direction—but human features were studied only in terms of their relations to the natural features. This view was more or less definitely stated by Wagner, Kirchhoff, and Neumann.

The outspoken emphasis on the division of the field into two sets of phenomena pointed logically to two opposite directions of development. That which came to have the more important effects can be traced from Ratzel through Semple to America where it had its logical ultimate expression in Barrow's statement of geography as the field of "human ecology," the mutual relations between man and the natural environment. This view has been echoed in England and also in Japan. Because of its great importance in this country we shall need to consider it later in detail.

The opposite conclusion-that man should be excluded entirely from geography-was urged much earlier, namely by Gerland in a long essay in 1887. Although this statement came "at a time that was certainly open to the ways of thought of natural science" and might wen have seemed the logical conclusion of the swing in geographic thought of the previous twenty years, it met with essentially no favourable response. It was clear to many, as Hermann Wagner observed in his lengthy critique, that if Gerland's apparently logical thesis were carried through to its necessary conclusions, so much would have to be excluded from geography that the remainder would not be recognizable as the field that had long carried that name. Few thought it necessary to give much attention to a thesis that would have thrown overboard the greater part of the geography of Humboldt as well as that of Ritter. Indeed it would be difficult to point with certainty to any effect of the controversy which Gerland introduced on the subsequent development of geographic thought in Germany; hence, while we will have occasion to return to it in other connections, we need not examine its claims here.

Historical Significance

The historical significance of Gerland's proposal was suggested a few years later by Wagner: the announcement of geography as a pure natural science came at the high point in the swing toward more intensive cultivation of physical geography that Peschel had started nearly two decades before. Although Gerland, as Schmidt remarks, was "simply consistent" with the course of this movement, none of the "morphologists" was willing to continue to *the* logical extreme. Indeed, it is possible that his arguments may have made some realize how far geography had been swung out of balance.

The counter-movement however had already set in, both in actual geographic work and in methodological considerations. A major event in the history of geography that was to have indirect influence on its methodological development was the publication, in 1882, of Ratzel's *"Anthropogeographie oder Grundzuge der Anwendung der Erdkunde auf* die *Geschichte"*. The term, anthropogeography, which has ever since been associated with Ratzel's name, reflects his background in the natural sciences, particularly zoology, but is misleading. It suggests the geography of man in terms of individuals and races, anthropological geography; whereas the major objects of Ratzel's concern were the works of man, particularly-thanks perhaps to Moritz Wagner the products of man's social life in relation to the earth.

Purpose of Ratzel's : Ratzel was therefore, like Ritter, a student trained primarily in the natural sciences for whom geography offered the connection between the natural sciences and the study of man. As we found in the case of Ritter, the fact that Ratzel's work was largely in human geography has led many to overlook his earlier background. Thus, Sauer's statement that he "got into geography through newspaper work" is less than half the story. According to Ratzel's own account it was his interest in nature that determined his career; he never departed from his early intention of "devoting himself to some kind of scientific study." He took his doctorate in zoology', geology, and comparative anatomy. Continuing his zoological research in southern France he sought to replenish his funds by sending travel letters to the *Kolnische Zeitung*, with the result that he was offered a position as scientific and travel reporter for the paper. Karl Andree advised him to use this opportunity to become a geographer, and though Ratzel at first planned to travel as zoologist he did ultimately, after six years of travel in Europe, the United States, and Mexico, make the shift. His first publications however were in zoology, and his work in geography included physical studies of the snow limit in the Alps.

Ratzel's purpose was to establish the study of human geography on a scientific basis. Far more successfully than Ritter, Schmidt concluded, he "remained true to the direction of the natural sciences." His great contribution was "to have brought this part of the geography of man, cultural geography into a scientific system by organization of the phenomena, and establishment of concepts and significant connections of the results obtained". Although he is commonly considered as primarily a follower of Ritter, in a sense his *Anthropogeographie* represents the first major systematic study of the geography of man as suggested in the culminating section of Humboldt's "physical geography." The fire of criticism that has been directed against his work ever since, does not lessen its importance in having demonstrated that the human, as well as nonhuman, aspects of geography could be subjected to systematic study, thereby leading to more reliable interpretations in regional geography.

In one major respect Ratzel's approach was different from that of Humboldt. Working purposely from the point of view of natural science, he organized his first *Anthropogeographie* largely in terms of the natural conditions of the earth, which he studied in their relations to human culture. Though this procedure was common among geographers concerned primarily with physical geography, others, such as Kirchhoff had studied human geography by the reverse method-by considering human conditions in relation to natural conditions. In his second volume, Ratzel himself largely reversed the process, but many of his followers, notably Semple in this country, maintained his earlier orientation and thereby established a procedure that dominated human geography for some decades-at least in this country. German geographers, however, changed their procedure much earlier, thanks particularly to the efforts of Hettner and Schluter. In the meantime, however, the fact that human geography was studied in terms of the influences of the natural environment on man led naturally to the concept of the field of geography as essentially the study of such relationships a concept commonly, but probably erroneously, ascribed to Ratzel himself. (In particular it is misleading to place Schluter in opposition to "von Richthofen, Ratzel, and their contemporaries" as Dickinson does.

In the year following the appearance of the first major study in systematic human geography, Richthofen presented, in his inaugural address at Leipzig, what came to be regarded as the programmatical statement of modern German geography. His earlier statement, in the concluding pages of the first volume on China, 1877, had been an attempt to restate the concept of geography that was common with Ritter and Humboldt, but, as Hettner notes, it placed bounds on geography that were broken in the book itself. In the same year, however, Marthe had clearly restored the areal or chorological principle as the dominant criteria of geography, using the terms chorography and chorology which various writers before him, including Peschel, had taken from the early Greek geographers. At the same time, he emphasized the study of distributions, which had formed a part of Humboldt's geography, and expressed geography in simple terms as the study of the Where of things. Marthe's concepts also found expression in a statement on the nature of geography—drawn up by a committee of which he was a member—adopted by the international Geographical Congress at Venice in 1882.

In his epoch-making address at Leipzig, then, Richthofen took over Marthe's chorological conception and made it a fundamental basis of his concept of geography. In this address, writes Plewe, Richthofen showed himself to be the actual one to inherit and carry forward the ideas of Humboldt and Ritter, and also, in a consistent sense, to fulfil Peschel. In contrast with the latter, he had the sound, unprejudiced historical sense to fit himself into the course of development and to determine his position exactly".

Richthofen had been trained as a geologist, and as a geographer he was primarily interested in geomorphology. Upon his election to the Academy of Sciences, he chose to enter the "physical-mathematical class" in contrast to "the historian Ritter" who had belonged to the "historical-philosophical class". Nevertheless, he recognized the value of Ritter's work in the development of geographic thought. If one reads his address against the background of the concepts of Humboldt and Ritter one observes how much less he was interested in the contrast between them than in the common elements of their concepts and work. He was able therefore, "following the precedent of Humboldt, to restore the close connection of geography to the natural sciences," and at the same time to restore Ritter's program to its place in geography.

We need not concern ourselves with Richthofen's specific statements; as Hettner later noted, though he set the direction of geographic thought for the future, he had not been successful in finding a sharp formulation for his concept of the field. But the essential thought is clear from his discussion : geography studies the differences of phenomena causally related in different parts of the earth surface. Likewise important, in the development of geographic thought, was Richthofen's exposition of the relation of systematic and regional geography to each other and to the field as a whole. The actual purpose of' systematic geography is to lead to an understanding of the causal relations of phenomena in areas, an understanding which may be expressed in principles that can be applied in the interpretation of individual regions, *i.e.,* chorology. (Richthofen distinguished between a first step, chorography, which is non-explanatory description, providing material for systematic geography, and chorology, a final step, the explanatory study of regions, based on systematic geography; but this separation has not been followed.

Systematic and Regional Geography

The major distinction in geography, between systematic and regional geography (chorology), is therefore not a difference in materials studied—- Richthofen recognized that it is not possible to limit the materials to be studied in geography-but a difference in method of study. Because of the heterogeneous character of these materials, they must be studied by classes, in systematic geography. On this basis he recognizes not two but three major groups: physical phenomena, biological phenomena, and human phenomena. Though man is of course a biological object, and to a certain degree may therefore be included with the studies other animals, the fact that his relations to the earth surface are governed by a host of factors not significant for other animals, requires an entirely different form of consideration. In the chorological studies, the causal relations between all the groups of phenomena unites them in a single unitary study. Not bound by any limited concept of science, Richthofen, like Humboldt, saw no objection to a single science considering different kinds of things that exist together and are bound together by causal connections.

Current Geography

Richthofen's program for geography and the fuller exposition of it which Hettner later contributed (first in 1895, but most fully in 1905) prepared the way for studies in regional geography interpreted in terms of the fruits of systematic geography., This was not new in geography, but rather a return to the method of Humboldt,. Though neglected, first by the followers of Ritter who studied regional geography with relatively little consideration of systematic geography, and later by the followers of Peschel who considered chiefly systematic geography, it had never been lost sight of Supan however felt that Richthofen's address, in giving more detailed attention to systematic geography, had subordinated regional geography, and, in 1889, urged its more frequent cultivation. Likewise Hettner vigorously encouraged the development of regional studies—by example, by his methodological statements, and through publication in the *Geographische Zeitschrift* which he founded in 1895. Nevertheless this aspect of geography hardly reached a position comparable with systematic, notably morphological, studies, until relatively recent times, particularly after the World War led to a focussing of German interests on the full character of European areas, notably those of Germany. The fact that Hettner is 59 often referred to as a promoter of regional geography should not lead one to suppose that he discouraged the development if systematic geography. Throughout the successive shifts in emphasis of the past forty-odd years he has consistently maintained the necessity for geographic research from both; of these points of view. In his first brief statement on the nature of geography, with which he introduced his *Geographische Zeitschrift*, in 1895, he commented on the neglect of systematic studies by the followers of Ritter, and credited the leadership of Peschel with having restored systematic physical studies in geography. His first detailed methodological study, of 1903, was limited to the fundamental concepts and principles of systematic physical geography in that of 1907 he considered systematic human geography, and in many others, since 1905, he has stressed the importance of systematic, or general, geography as a coordinate part of the field especially. Finally', in more recent years, he has himself presented detailed systematic treatises covering different categories of geographic features, first on surface forms, then on climate, and finally, a four-volume work covering the whole of systematic geography . Human geography however is limited to less than a hundred pages of the fourth volume, since Hettner planned to devote a separate volume to this part of the field. Geographers of all countries will hope that nothing will prevent the publication of this volume for which the manuscript is essentially completed.

In his first discussions of the relation of these two types of geographic work, Hettner introduced a somewhat unusual terminology designed to emphasize that there was no sharp separation between them. In a regional study of any extensive area it is necessary to study the notable variations in the individual geographic features systematically. On the other hand, the systematic study of a particular category of geographic features it not made with reference solely to that category, but rather in terms of its chorological relations to one or more other features—i.e., their relations as each varies in different areas. He wished to emphasize, that is, the distinction that Humboldt had clearly drawn, but which nevertheless had often been lost sight of, between studies in systematic geography and those of the special systematic sciences studying the same objects on the relation of Hettner's views of geography to those of Humboldt.

In the German literature of the nineteenth century the terms introduced 'by Varenius, "general" and "special geography" had in considerable part been replaced-in the desire to eliminate words of foreign' origin-by *"allgemeine Erdkunde"* and *"Landerkunde."* Not only did these suggest too great a cleavage between the two types of geographic study bat the term "general science of the earth" expressed a very different concept from that commonly associated with the word "geography." It was forgotten, as Penck has recently observed, that *"Erdkunde"* was "simply the germanization of the word geography. One took it literally and derived therefrom the task of studying the earth as a whole." Gerland, under whom Hettner had studied, had attempted to follow this reasoning to its ultimate conclusion and even Richthofen was apparently confused by it. One consequence, Hettner felt, was that geographers were influenced to make systematic studies of the earth as a whole in which the chorological point of view was lost. Geography became even more dualistic in nature than Wagner had recognized; studies in systematic geography did not produce general concepts and principles concerning the interrelation of phenomena that were needed for regional geographic studies. To overcome this separation of the field into two entirely distinct types of work, Hettner regarded it as necessary to make the chorological concept dominate—as in Humboldt's work-in systematic as well as in regional studies, To emphasize this viewpoint he placed at the head of his outline of systematic or general geography -drawn up in 1889, two years after the publication of Gerland's proposal the title *"Vergleichendc Landerkunde."* The relation of this form of geographic study to *Landerkunde* proper, or regional geography, is clearly explained in his studies of 1895 and 1898. Although Hettner dropped this term in the methodological studies in which he most fully developed the relation of the two kinds of study, later, when he ultimately fulfilled the outline written more than forty years before, he maintained the original title.

Certain Features

Before leaving this period of the late nineteenth century, we must note certain characteristics of the geography of the classical and pre-classical periods that, for a time at least, almost dropped out of sight. One of these is the concept of unity or *Gilllzheit* in geography. In his survey of the development of this concept in the history of geography, Burger is able to present but little acceptable evidence that it was important in German geography during this period. Relatively few German students appeared to have been concerned with Ritter's concept of the earth as an organic Whole, though Ratzel may have echoed this thought in certain connections, and it was a favourite concept of the great French geographer, Vidal de la Blache. Gerland' effective argument against this concept reflects an increasing demand in science for direct accurate description, in place of either mystical assumptions or misleading analogies. The earth, he concluded, was not an "organism" but simply a "complex of cosmic matter".

To be sure, German geographers of the late nineteenth century continued to study the mutual relations between different phenomena in areas; Richthofen saw the unity of the field as resulting from the causal interconnections of different kinds of phenomena, and Hettner, as well as others, emphasized the importance in regional geography of presenting not merely some of the characteristics of an area, but the total character, as determined by the interrelated combination of all significant features. But this total character is not thought of as a "unity," an *"Einheit,"* or *"Ganzheit"* in the sense of the philosophy of early writers, including Humboldt and Ritter.

Likewise, one does not find during this period the concept of the individual region as a unit in itself, a "whole," or an "organism." Both of these aspects of Unity and Wholeness have been returned to geography in recent decades, chiefly since the World

War. Its belated development makes it questionable to consider it as a result of the "return to Ritter," of which both Richthofen and Hettner spoke in a very limited sense. Certainly neither "Richthoten nor Hettner ever supported the concepts; the latter, as we shall see, has frequently challenged them. He himself suggests that the concept might be traced from Ritter's teleological concepts, presumably through Ratzel. It was evidently from Ratzel that Vidal de la Blache took his concept of terrestrial unity, of the "earth organism"; though Vidal erred in supposing it was new in geography. But it is also possible that the notable development of these concepts in recent years is not so much a product of the ideas of Ritter and Humboldt, but rather a repetition of a similar cultural phenomenon-the introduction into geography of general philosophical concepts of a particular time and country. In the post-War atmosphere of Germany. "Unity" and "Totality" are powerful concepts.

In the scientific temper of the late nineteenth century there was apparently little place for the consideration of the aesthetic character of landscapes that Humboldt had so notably developed. Burger cites as sole exceptions of note, studies by Ratzel, Oppel, and Wimmer. Both of the latter continued Humboldt's usage of the term *"Landschaft"* in the-sense of the visual scene.

In sum, we may say that in the latter half of the nineteenth century, under the influence of the development of the specialized natural sciences, geography for a time appeared to be changing into a field quite different in character from that which Humboldt and Ritter had inherited and developed: The emphasis on systematic studies appeared to divide geography into two halves, one a natural science, the other a social science, united only in a study of regions that hardly appeared to be a science at all, in the sense in which the time conceived that term. By the end of the period, however, reaction had set in, so that the direction of geography was again essentially that which it had been before. At the turn of the century, the purposes of geography corresponded in major degree, as Richthofen stated in 1903, "in content and methodology, to the conception which Humboldt had given it". Geography then, as throughout most of its history, was concerned, as Hettner put it, to study the areas of the earth *(Erdraume)* according to their causally related differences-the science of areal differentiation of the earth surface.

Hettner's Approach : Although Hettner found in 1905 that German geographers were generally coming to accept Richthofen's statement of the field of geography, he, as well as others, felt that Richthofen "had not been entirely successful ,in finding the sharp methodological expression for his concept" that he had not followed it through consistently, and, later, in his address of 1903, had somewhat obscured the issue. Hettner therefore set himself the task of providing a sound methodological exposition, not of some concept of geography which he deduced in his own mind, but of the concept which seemed most closely to represent the field of geography as it had developed hitherto. He was therefore not merely standing on the shoulders of Richthofen, but likewise, both indirectly and directly, on those of Ritter and Humboldt and their predecessors of the pre-classical period. His dependence on Humboldt can be noted in countless passages; he himself has indicated it in many. Doring concluded that in Hettner's synthesis of the geographic thought of the two founders of modern geography, Humboldt stands not beside but before Ritter.

Throughout the long series of methodological discussions which Hettner has published from 1905 to the present time-with further studies, one may add, ready for publication-he has consistently endeavoured to express the concept of geography as the outgrowth of its historical development. In that sense they represent the culmination of our study of the historical development of geographic thought. Since his purpose in these studies, however, has been to provide the logical basis for the concept historically evolved, we will examine his views in the following section.

Our historical survey may be briefly' concluded. The viewpoint which Richthofen had stated in 1883, as interpreted by Penck, Schluter, and above :111 by Hettner, came to be widely accepted, so that German geography in the first part of the twentieth century has been marked by a greater degree of unity of fundamental concepts than ever before. In particular, Hettner's methodological discussions have come to be regarded as "classics" in geography which no German scholar would ignore in any consideration of the methodology of our field. Solch wrote in 1924 that "almost all the scientific geographers of Germany have arrived at similar conceptions, thanks not least to Hettner's work".

Likewise the methodological studies of Hettner and others of his contemporaries have had a notable influence on the concepts of geographers of other countries. Among the geographers of continental European countries outside of Germany who have expressed similar views on the nature of geography we may list the following: Berg in Russia and his student Marcus in Estonia Grano in Finland; Arstal in Norway (according to Braun) and Helge Nelson in Sweden (as quoted by DeGeer, who on the other hand expresses a concept similar to that of Marthe; Michotte in Belgium; and in Italy, Giannitrapani and Marinelli (according to Solch) and Almagia. Both Chisholm and Herbertson attempted to introduce Hettner's views in Great Britain. In Japan, a number of geographers today follow Hettner, according to Inouye—including Komaki and Watanuki and, evidently, the reporter himself. The chorographical concept was first introduced into this country by Fenneman and is to be faun in one of the later discussions of W. M. Davis, in marked contrast to his earlier, better known statement; further it may be discerned in the work of many others, notably of Wellington Jones. Nevertheless it was largely overlooked by American geographers until expounded with great effectiveness by Sauer a little over a decade ago; since then it has become the general concept of the field for a large number, if not the majority, of the active research workers, in the United States.

On the other hand, in countries outside of Germany, there is no such general agreement among geographers as to the general nature and scope of their field of study as has characterize German geography of the past generation. In most countries indeed there is marked divergence and conflict of opinion. It is significant to note therefore that geographers outside of Germany have shown but little interest in studying the methodology of their field, as Stamp observes in a recent survey—though it should not be supposed that they do not frequently talk about it and publish their individual views.

French Geography

French geography presents an exception, in the relatively high degree of unity that was established by the predominating influence of Vidal de la Blache. But important as the influence of Vidal, Brunhes, and Demangeon has been in the development of regional geography, it has resulted from their Works, and those of their students, rather than from studies of the nature of geography. Indeed if one read only what these writers say about geography, as distinct from what they have done in geography, one would have to place them simply as modifiers of Ratzel. The most thorough, and in many respects a highly valuable, analysis of geography from this point of view is Vallaux, *Les sciences geographiques*, which will be referred to frequently in later sections of this paper. It must be expressly stated, however; that no adequate attempt has been made here to survey the current thought in French geography; the reader will find some references in the recent survey of work in that country by' Musset.

British geographers have shown perhaps the least interest in attempting to determine the nature and scope of their field. Chisholm's inaugural lecture of 1908, based largely on Hettner, appears to have had little influence. Both Solch and Huender (of Utrecht) have noted that different English geographers express widely differing views, though nearly all are subject to the "environmentalist conception". At least one English geographer, Dickinson, has expressed distress at the lack of coherence in geographic work in England. Roxby's presidential address in 1930 examined the concepts of earlier German geographers and the interpretation of them by Vidal and Brunhes, to arrive at a statement very similar to that of Barrows; but German geographers since Ratzel are confined to a footnote.

The following year Mackinder sought to state the unity of geography in terms of natural regions based primarily on the hydrosphere rather than on the lithosphere-a conception that no doubt owes much to Herbertson. Bryan's book on "Cultural Landscapes" shows no acquaintance with the developments in Germany since Ratzel. On the other hand the work of the Passarge school is discussed at length in Unstead's study of regional geography and Stamp's survey of economic geography considers American, French, and German points of view. In the last two or three years both Crowe and Dickinson have given critical attention to more recent German views—in part, apparently, as a result of the stimulus from this country.

That American geographers have been more given to discussing and arguing over the nature of their field than to studying the problem is well-known. Serious scholarly studies of this problem, including a due consideration of the findings of previous students, are limited chiefly to two or three presidential addresses, already referred to, and the well-known discussions of Carl Sauer. Insofar as Sauer's methodology is derived from German writers it depends largely on Schluter. Though he recognizes Hettner's methodological studies as "at their best perhaps the most valuable appraisals of what geographers are trying to do" he appears to have otherwise largely ignored them, perhaps because of a misunderstanding of Hettner that he appears to have taken over from Schluter.

Though the majority of American geographers may not have written on this subject they nevertheless have pronounced, but divergent, convictions as to the proper definition and scope of geography. Categorical criticism of specific papers on the grounds that they are "not geography" are common, and highly debatable views of the nature of the subject are published, of all places, in college textbooks. One feels justified in asking these critics to give us the fuller benefit of their Judgement in thorough scholarly studies of the nature of geography.

The geographers in Germany and the large number in other countries who agree with them in considering geography as a chorological science concerned with studying the areal differentiation of the world, do not suppose that they present a new concept of an old field. On the contrary, as many, including Hettner and Sauer, have noted, this concept may be derived from the work of the earliest geographers, from Herodotus and Strabo. Modern geography, Sauer concludes, is "the modern expression of the most ancient geography" it is appropriate therefore to use the term common in ancient geography (and revived by Marthe and Richthofen) , namely, chorology, the science of regions.

2

History and Geography

Incorporation of Geography in History : The first of three major problems in geography today is presented by the interlacing relations of geography and history. To what extent should these two fields include each other? The recognition of the importance of the similarity of geography and history as integrating sciences should not lead to a confusion of their distinctive functions. If each of them may be thought of as cutting cross-sections through the systematic sciences, these cross-sections are in separate dimensions. The fact that an illustration of this relation requires a four-dimensional space reflects the extraordinary difficulty of producing a reasonable synthesis of the two points of view and the two bodies of materials—a difficulty which may well, as Huntington suggests, "make thinkers pause where elementary school teachers plunge without fear". (The confusion of purposes produced by the "fusion" of geography and history in instruction was ably treated a century ago by Bucher).

Nevertheless, however great the difficulties may be in relating the two fields, one can hardly question Huntington's statement that "a rational understanding of history requires a good knowledge of the changing physical background upon which the historical events occur." Since historians are usually not equipped with such geographical knowledge their attempts at synthesis have not, he finds, been highly successful.

Of the geographers who have worked in this field, Huntington observes that most "have selected only those historical items which illustrate their special points of view" and modestly pleads guilty to his share in this offence. He evidently does not find this objection in the work he was reviewing, Griffith Taylor's *Environment and Nation*, but nevertheless concludes that its value is "that it calls our attention to the fact that. . . we are at the very beginning of our task. We have only reached the point of accumulating materials for a foundation on which our successors may build a fair and seemly structure." Critical reviews by other geographers would suggest that in such an extremely complicated problem it is more likely that sound foundation materials will be obtained from studies more localized in time and space than the entire history of the European continent. Any students who have the hardihood to tackle this problem may well be urged to seek their materials not in any of the better known general studies but in the large number of less ambitious regional studies which may be found in the German and French periodical literature.

Geographers, in particular, who wish to consider this type of problem, will find a special reason for confining themselves to a limited period or area, or both. In view of much of the work that has been done by geographers in this field, one is tempted to add to Huntington's statement of what is needed for a rational understanding of history a sound knowledge of the historical events. The student trained as a geographer will do well if he can acquire this knowledge for a very limited area.

The discussion has its place in this paper because the subject considered has been studied primarily by geographers, commonly under the title of "historical geography." Huntington's statements, however, clearly indicate that the focus of attention is not on geography but on history-it is the phenomena of history which are studied in terms of their geographic aspects. Logically it appear clear that this is a problem within the field of history. One of the leading American students of the problem, the historian Turner, indicates this most clearly in describing himself "not as a student of a region but of a process". Probably most geographers would now agree that the geographic interpretation of history is logically a part of history, even if the contributions have hitherto been largely those of geographers.

Incorporation of History in Geography : Within the field of geography itself what consideration should be given to the sequence of historical development? Some geographers insist that in order to maintain the essential point of view of geography the consideration of phenomena in their spatial relations-any consideration of time relations must be secondary and merely supplementary. Others however urge that the geographer is primarily concerned with the development of, or changes in, the phenomena which he studies; time relations therefore become of major importance.

The thesis that the geographer is primarily concerned with the changing character of an area, has been vigorously proclaimed in Germany by Spethmann, under the title of *Dynamische Erdkunde*. In contrast with the work of other German geographers which he characterizes as "static," he proposes the "new dynamic view" of the subject. Actually, as Gradmann notes, there is nothing new in Spethmann's idea; his predecessors did not have a static concept of geography, he is merely using a *Modewort* (a word ill vogue).

Spethmann's proposals aroused a great stir in German geographical circles, not only because of the vigor with which they were put forth, but also because of the unfortunate personal attacks that accompanied and followed them. While some received it with praise, others, notably Hettner, Philippson, and Gradmann were highly critical. Spethmann filled a second book with replies to his critics, but few who read Gradmann's dignified review will care to wade through the former's personal complaints. On the other hand, credit must be given to Spethmann for having put his ideas into a concrete example, namely in his three-volume study of the Ruhr area, in which he follows the successive changes in the development from prehistoric times down to the present. In the two volumes of this study that are required to bring it up to the present time the reader (some may prefer to rely on my review) will find ample evidence of the way in which a geographer is led out of his field when he permits "genetic explanation to become historical narrative".

In this country, Sauer has specifically insisted upon the use of the historical method in geography; since present "cultural landscapes" are to be studied in terms of their development from the original "natural landscapes," the study must be oriented on development. This orientation apparently requires that we begin at the beginning. Lest this force us to start with the origin of the earth, he suggests that "geography dissociates itself from geology at the point of the introduction of man into the areal scene". Starting at that point with a reconstruction of the original "natural landscape," the student of geography will work through the various different transformations in the landscape caused by man (and presumably also those caused by nature) until he arrives at the present "cultural landscape."

Parenthetically it must be noted that geologists can hardly be expected to accept the line of demarcation suggested. The development of the earth's crust, the formation of rocks, and the deposition of fossils which the geologist studies, did not cease at that unknown time when man arrived upon the scene. On the contrary, if man's activities in any area today cause changes in the form or character of rock formations-slumping of beds of shale under heavy buildings in Cincinnati, decrease in water content of sandstone formations because of artesian wells in Minneapolis-the geologist is concerned to study those.

Study of Region

More important are the practical consequences of Sauer's thesis. Evidently any study of a region must include, indeed begin with, a study of its entire previous geography, chronologically arranged. In his own study of "Site and Culture at San Fernando de Velicata", the emphasis on the chronological order is so marked that it is a fair question whether the study is not to be considered as history rather than geography. In later studies Sauer, logically enough, has gone farther back into prehistoric times and it becomes difficult to dissociate the work from anthropology. Similarly Trewartha, following Sauer's precepts, presents as the first chapter in the geography of the Driftless Hill Land (of the inner basin of the Upper Mississippi), a chronological treatment of some nine or more French trading posts, each occupied for an average of but little over three years—a study that would be very difficult to distinguish from a study in history. These comments are not intended in any sense as a criticism of the studies concerned, nor would the, writer object to any who wish to make such studies. The question is whether geographers who wish to study the present geography of a region are required first to produce works for which few of us are technically prepared and which can hardly be distinguished from those of other fields.

Although no American student, so far as I can find, has presented a complete illustration of the thesis that Sauer propounded more than a decade ago, Broek's doctoral dissertation at the University of Utrecht, on the Santa Clara Valley, might possibly be considered as in some degree a product of Sauer' concepts, though it follows Grano also in considerable degree. Broek's introductory statements would indicate that a regional study should consider the "landscape" of every period from the earliest to the latest, but evidently one may recognize differences in degree of significance: in any case, he omitted landscape changes before white settlement, starting with the "primitive landscape" that the first Europeans found. The major part of the work considered the changes shown in successive landscapes of six historical periods." In each of these periods it was of course, necessary for him to follow changes in each of various important elements in the landscape-as' Creutzburg has noted, a landscape is not a whole that grows as a whole, but only as each part grows, so that "the history of a landscape becomes the histories of its parts". Consequently we do not get a full picture of the landscape in anyone of these periods, but rather a genetic study of various parts. Finally, the reader must be prepared himself to fit these into the relatively limited section devoted to the present landscapes. Unquestionably, a thorough study of this work would give the reader a better understanding of the present landscape of the Santa Clara Valley than could a treatment that considered only the present without looking to the past. But, needless to say, the latter is not the only, alternative, if it is even an actual alternative; the so-called "static" method of regional study does not limit itself to present conditions. The question here is whether we would not have an even clearer picture if the study had been organized entirely on the present landscape, and in the analysis of each of its parts had utilized the appropriate material that was presented in the historical chapters.

It follows that the student of any area must learn far more of its historical geography-and of its history in the broadest sense than he needs to impart to the reader, in the same way we may be permitted to suggest that an historian of any area needs to become familiar with its geography though he may include but little of that in his historical writing. It is not always possible to know during the course of a study, which facts from its past geography will be significant for the present. The necessary selection can be made with greatest ease for reader as well as author if the facts of past periods are presented at the point in the writing where they are significant to the present.

The logical extreme of the historical point of view is to be found in Leighly's suggestion that the cultural geographer should follow the trail of the cultural historian. "This above all that one emphasizes the essential time bond of culture rather than its looser place bond". Pfeifer's comment that "thereby the antithesis of environmentalism has been reached" no doubt correctly interprets the course, of thought that led to Leighly's statement, but is it not at the same time, one must ask, the antithesis of geography?

Historical Methods

The historical method appears particularly paradoxical in studies that claim to limit geography to visible or perceptible objects. Insofar as this concept has any merit, that would seem to be completely lost in studies in which the major part of the work consists of descriptions of things the author ascertained only from historical records, on the basis of which he ultimately emerges with the final product—namely that which he could see from the start. Indeed, as Hettner pointed out, the majority of facts that the geographer can directly observe are by nature static, the processes studied in a "dynamic" regional study must be deduced hypothetically from the static facts-or taken down from historical records.

Following a similar line of reasoning, Finch recommends that the conditions of past periods should be drawn upon only where necessary to interpret present features that represent relict forms of the past. This may require the reconstruction of the geography of past periods, but such descriptions, while not extraneous, are but supplementary, "in a sense they are asides." Far from commencing with the study of the earliest periods, Finch would commence with that aspect of the area on which the writer can be positive, the present scene. Much the same view has long been held by such students as Hettner, Hassinger, and Unstead, to mention but a few.

In opposition to these practical arguments is a logical argument, based on a particular concept of geography, that appears to require the geographer to study the origin and development of an area in the time sequence. If the particular objects of the science of geography are the concrete "landscapes" of the world, then geography must study these objects, not only in terms of their present form and function, but also in terms of their genesis-their origin and evolution. Though a consideration of the concept itself must be reserved for a later section, we may consider here its relation to the present problem.

To be fully consistent, the study of the origin of any landscape would apparently require us to study the origin of the continents, if not of the earth itself. Sauer, as we noted, avoids this by somewhat arbitrarily cutting the sequence at the point where man entered the scene, assigning to the geologist the study of the development of the "natural landscape." (This would require more than a geologist, and the total of studies by representatives from geology, climatology, plant ecology, etc., would not provide a study of the development of the "natural landscape"—as Sauer uses the term. Someone else must put together the results of the work of these students.) The geographer is to study the development or the evolution of the "natural landscape" into the present "cultural landscape." This expression has been repeated so often, on both sides of the Atlantic, that one tends to read it without questioning its import. How does a "landscape" by whatever definition—develop or evolve? The frequent analogy of "landscape" or region with an organism suggests, as the word "evolve" itself does, that the "landscape" is something that develops or grows as a unit from forces within itself. If, however, one clears one's mind of misleading analogies and dubious concepts and considers the question in straightforward terms one sees at once that the process of change that we may observe in either the "landscape" or the region is very different from the process of organic development or growth. This conclusion may seem so obvious to the reader as to require no demonstration, but the fact that it is so often ignored requires us to demonstrate it in some detail.

If the term "landscape" is used to' mean something less than all that is in the area, in particular as exclusively the material features in the area, then it would be painfully ridiculous to say that a "landscape" sheds its forests, produces ploughed fields, or grows houses and factories. These elements of the "cultural landscape" are produced in the "natural landscape" by forces that by definition, are not part of the "landscape"—namely the forces directed by human will. In other words, though we might say that the earth surface as a whole has evolved cultural landscapes out of natural landscapes, the "landscape" as such-either the real visible landscape surface, or the collection of material objects of the earth's surface-has not *developed* but has been *produced* by factors not included in it.

There are, however, :several critically important respects in which the processes of change in areas differ from those which we associate with organic development. In his discussion with Dickinson, Crowe observes that the changes that take place in cultural forms and human societies have "none of the inner necessity characteristic of organic evolution. The generation of new forms is open to human choice and initiative, it is a matter of Invention and Organization. . . . Close analogy with living matter, or even with geomorphology, is on the highway to falsehood. More important, perhaps; is the objection raised by Creutzburg, that there is no unity of growth in the development of an area (or of a landscape), there is only a summation of partially related, but partially independent changes within it. The forces producing these changes-both the will and energy of the individual inhabitants and the changes in individual natural' elements-do not form a unit force, but merely a summation of more or less independent, often conflicting, forces. Broek therefore rejected the term "development" and referred instead simply to "changes" in an area or landscape.

Previous Objections

The previous objections apply to the consideration of the entire area of the earth's surface. When we consider a particular areal unit, or "individual landscape," however determined, there is the further major objection that the changes within it are not necessarily produced by its own forces, but may be caused by forces entering from other areas. The landscape of the Lower Mississippi flood-lands was in part produced-and may at any time be altered by conditions in the Northern Appalachians. Bluntschli has described how the "harmonious landscape" of the Amazon lowlands has been disrupted by forces and materials introduced from lands thousands of miles away. Indeed Schluter has argued that most changes in areas are caused by the movement of peoples from other areas. To compare such changes with the development or growth of an organism, or even with the inanimate development of crystals or of a residual soil, is to mislead ourselves. Likewise it would be misleading to suppose that these changes introduced from the outside were controlled by some unifying force, either within or without the area. The natural landscape is not converted into a cultural landscape as an artist applies extraneous materials to a canvas to "develop" a unit picture, for the changes are not the work of one artist, or of an organized group of artists, but merely of a collection of somewhat independent natural and human forces.

The various forces that alter the landscape of an area, whether they are internal or external, recognize no common limits to the area. It follows therefore that, in whatever manner we may consider a particular area as a definite unit, that unity can be established only as of a given time. A unit area today was probably not a unit area in an earlier period, and its present unity may be disrupted tomorrow by changes that may so alter portions of it as to include them in a neighbouring areal unit. Any study of the development of the cultural landscape of an area, Broek therefore concludes, is legitimate only if we remember that the area considered through a sequence of periods is an arbitrary unit. Whatever interest there may be in studying the combination of processes of changes in an arbitrary unit of area, there can be no logical requirement that geography must make such studies. On the contrary, as Broek emphasizes, this thesis is in principle, a survival of the idea of environmental control." Underlying it is the assumption—commonly unmentioned, and even denied-that the unit natural area (or "landscape") will remain a unit through all changes made by man and so lead to a unit "cultural landscape." In reality, however, observation shows that "cultural landscape regions are often a combination of parts of different natural regions" and we may view "cultural landscape regions as expanding and contracting areas, perhaps even as migrating areas".

From these conclusions it follows that, not only is there no logical necessity for the student of a region to examine each of various stages of "development" of the area, but also that the presentation of such a sequence of historical cross-sections will not necessarily provide the best means of understanding the present situation. Since whatever degree of unity involved is to be found only in the present area, we may more readily arrive at an interpretation of the present by analyzing the features found in it, utilizing the facts of the past only to explain, for each of the present features separately, the factors that produced it.

Whereas most supporters of the "dynamic" method of regional study defend it as a means to a proper understanding of present iandscape. Stanley Dodge inverts the argument and denies scientific standing to the study of the present alone. "Much of the writing of geography treats of regions as static, as being, whereas it is the becoming which is important. What processes have shaped regions, what processes are continuing the elusive transformations, and what are the trends for the future-these are the questions which excite the field geographers". These questions, Dodge adds in correspondence, present real problems to the geographer; merely to learn what a region is like, is no problem worthy of a scientist.

The particular problems that may excite the individual field geographer are undoubtedly of first importance to him, but that is a question of subjective reaction varying for different individuals. If, however, one examines the question of what "is important" objectively, one must ask what importance becoming can have, if the state, of being is unimportant. In any branch of science, the study of becoming is significant in order to explain that which is (or was) ; it is therefore of secondary importance, Even in that field which seems to be largely concerned with the processes of development, history, the purpose, as Kroeber shows, is not to be found in the sequence of changes in time; those are studied in order to explain the integration of phenomena of any particular time.

We may say categorically that what is in a region, its being, is important if anything is important. It is true that this does not present the geographer with a single clear-cut "problem," such as those which the students of the systematic sciences find in examining individual types of relationships, but rather a much more complex general problem corresponding to the problem of the historian in interpreting the associated phenomena, say of English History in the reign of Elizabeth. To comprehend fully the present geography of a region-that is to know the facts and be able to interpret them—is by no means too small a problem for any geographer, nor need we be concerned lest we shall shortly exhaust our field of problems. "There is no region of the world," Penck notes, "whose geographical investigation can be considered as closed" though a generation has passed since Penck made this statement, his further comments on the continuous growth of new problems indicate that he would repeat it today. Individual students will of course study whatever problems interest them, but there is no need for geography as a whole to look for new problems, as long as it has the present world before it in all its manifold and complex diversity.

Concluding Remarks

We may conclude therefore that while the interpretation of individual features in the geography of a region will often require the student to reach back into the geography of past periods, it is not necessary that the geography of a region be studied in terms of historical development. On the contrary, unless the student undertakes the ambitious task of presenting simultaneously both the history and the geography of a region he will need to distinguish clearly between the two points of view that which studies the associations of phenomena in terms of time, and that which studies their associations in terms of place. If the chorological manner of viewing things is the guiding principle of geography, the chronological organization of a study, even though of a "geographic region," is evidence of the essentially historical character of the work.

The relation of geography to the facto_ of time, Hettner concludes, is not that of a contrast with geology, according to which geography is limited to human times, historic times, or even to the present as compared with the past. Rather, geography is a field for which "time in general steps into the background." In contrast with all the historical sciences, including historical geology, "geography does not follow the course of time as such-to be sure this methodological rule is still often overlooked-but lays a limited cross-section through a particular point of time and draws on temporal development only to explain the situation in the time chosen." This one particular time must be short enough so that within it there are no major changes disrupting the cross-section. To be sure, the interpretation of the present regional geography will often require a consideration of past conditions. Geography requires the genetic concept but it may not become history.

The Concept

A coordinate combination of the two different points of view under which phenomena may be integrated the geographic and the historical—is not geography or history but both. Whether or not such a combination is humanly possible we need not further discuss, at least until someone has demonstrated its possibility in practice. On the other hand, the fact that time, in itself, is not a factor in geography, that geography studies the integration of phenomena in areas under the assumption of fixed time, does not limit geography to the present. It may strike a cross-section through reality at any point of time. Whereas we commonly assume that the term "geography," when used without qualification, refers to a cross-section through the present, we may use the term "historical geography" for an exactly similar cross-section through any previous point of time.

Historical geography, therefore, is not a branch of geography, comparable to economic or political geography. Neither is it the geography of history, nor the history of geography. It is rather another geography, complete in itself, with all its branches. Consequently, Hettner observes, "an historical geography of any region is theoretically possible for every period of its history and is to be written separately for each period; there is not one but many historical geographies".

In other words the significance of the word "historical" in this term is not to imply a direct connection with the field of history, but is used in the sense of "historical" as meaning that of the past. There is involved, to be sure, a practical relation to the field of history, in that the data for a study in historical geography must be obtained in large part from "historical" materials. Further, the historian of a past period may be interested in the historical geography of that period, insofar as knowledge of that is necessary of an understanding of its history, but this is no different from the interest which an historian of the present period may have in present geography.

We may make the parallel situation of the two fields complete by saying that, while history is concerned with the integration of phenomena in periods of time, it must recognize more or less separate histories for each major area of the world. Likewise geography, integrating phenomena in areas, recognizes separate geographies of each period of time. Geography, in the sense of present-day geography, therefore corresponds to the history of one's own country; historical (past) geographies, to the histories of other countries. A practical difference results from the fact that, while the study of foreign histories permits the use of the same methods as the history of one's own country (except for differences in language), the study of past geographies does not permit of the direct field observations of present geography. Further, there is a difference in interest for reasons that need not here concern us, the people of any time and place have a greater intellectual interest in the history of other countries-both present and past-than in the past geography of their own country.

Geography of Past Period

That historical geography is to be considered simply as the geography of past periods' is a view on which there is perhaps more agreement among geographers than on almost any other question of definition in our field. It was stated early by Kant, in the distinction between "the geography of the present" and "old geography", and later by Marthe and Wimmer. It has been most clearly expounded by Hettner, briefly in his first statement of 1895, and more fully some years later. Similar views have been stated since by so many geographers of different schools in various countries that we may be justified in considering it as established in geographic thought. In this country, in particular, the nature of this field' and the methods by which it may be studied have recently been discussed and illustrated with a detailed example, by Ralph H. Brown. (For similar expressions concerning historical geography, see Hassinger, Brunhes, Unstead, the group of British geographers and historians reported in. Gilbert, Barrows, Sauer, and others noted in Hartshorne. Vogel discusses the German literature in this field, with extensive bibliography).

Hettner and Brown : Hettner observed, and Brown has demonstrated, that it is not necessary for historical geography to begin with the earliest period in the history of a region. As geography, it does not begin, proceed and end, according to a time sequence, but rather considers the geography of any period the writer desires, exactly as one would consider the present geography. Nevertheless most students in this field, under the influence of the historical point of view, have felt it necessary first to attempt to reconstruct the pre-historical landscape and then those of later periods. As Brown has pointed out, however, there are very few areas of the world for which we have anything like reliable and adequate data either for the pre-historical landscape or for those of early periods of history. The imaginative reconstructions which can be derived from a consideration of the present cultural landscape and from the meager and highly unreliable descriptions of early explorers and settlers may be very interesting, but are of dubious scientific value. Sauer's suggestion that "the major datum line from which human 'deformations' are to be measured" is to be attained from "a reconstruction of the natural 'original' vegetation," *i.e.*, that existing "before man entered on the scene" whenever that may have been-is, simply to recommend the unattainable. Indeed more often we must arrive at the "datum line" by a consideration of the historical record of the deformations.

In order, therefore, to work with more reliable materials, Brown deliberately ignored earlier periods for which there is little reliable material, and selected the first period in the history of his area for which he found current material of sufficient geographical quality and reliability to justify a scientific reconstruction of the geography of the time.

It is true that Brown's method requires him to depend largely on historical sources, *i.e.*, the records- of the past; and to some extent therefore to utilize historical methods. This would appear to justify Hettner's conclusion that "the cultivation of historical geography should lie mostly in the hands of historians". But Brown has demonstrated that to check and interpret the historical records of geographical material requires the knowledge and abilities of the trained geographer. He has illustrated Kant's observation that a knowledge of the geography of the present provides the means by which historical records may be used to interpret the geography of the past. In doing this he has not followed the historical development as a purpose in itself, but rather has used that for the purposes of geography. Consequently his production-the ultimate test—is essentially different in character from that which any historian could or would have produced.

There is a further consideration that might move the regional geographer to make a detailed study of the historical geography of his area, at one or more past periods, before presenting his study of the present. The geography of certain areas is so different from what one would expect from present conditions and has perhaps been determined very largely by the geography of some particular period of the past-rather than by a consistent development of fairly permanent conditions-that the present can most readily be understood if the historical geography of that particular period is first portrayed. An urban study of Venice would offer an extreme example -a full interpretation of the geography of Venice in the fifteenth century is of first importance for an understanding of the present city. Similarly, in beginning a study of the political geography of the Mid-Danube lands in 1937, the writer found that so much of the situation of that time (the same would still apply) had been conditioned by the political geography of the nineteenth century that it was desirable to present first a fairly detailed study of the political geography of the prewar Austro-Hungarian empire, There seemed no such need, however, to present in that way the situation of earlier periods, though the detailed analysis required, at certain points, an examination of particular features of much earlier centuries.

There remains, finally, a form of study in historical geography which much more closely approaches history. If a geographer studies the historical geographies of a single region—or of a part of a region-at two or more different times, he will of course be concerned with the-differences among the several pictures. This, I take it, is the type of study which particularly interests S. D. Dodge, and, though marked by a strong historical bent-the study of changes-it would appear to have its place as a form of comparative historical geography. (Compare also Volz' concept of historical rhythm in geography, and the discussion by Burger).

Since the comparison is that of the geography as found in cross-sections through several periods of time, such a study would presumably be organized with major divisions for each of those cross-sections, but within each of these major divisions a geographical study will not follow the chronological timetable but will be organized chorologically. A specific example is offered by Robert B. Hall's study of the Tokkaido road and region, in Japan, which presents the geography of that section of land at three different periods of time. Solch has commented on the contrast, between studies designed to present a temporal cross-section of geography in the past and studies focussed on the historical events and economic development that have changed areas, that is to be found in the chapters, written by different authors, in the "Historical Geography of England before 1800," recently published by Darby.

Theoretically one might construct an unlimited number of separate historical geographies of any region, and if these could be compared in rapid sequence one would have a motion picture of the geography of an area from the earliest times to the present. In practice however this is utterly impossible hence indeed the separation of history and geography. We can either compare a relatively small number of cross-sections of the regional geography of past periods, or we can follow the detailed process of development' of particular individual features. But the study of the development of any particular feature is a part of the systematic study of that type of phenomena; depending upon the point of view under which the process is studied, it is either a study in the systematic science concerned with that type of phenomenon, or it is a study in systematic history.

A complete combination of history and geography, in conclusion, would represent the complete integration of the areal and temporal variations of the world-it would be reality. But we, have no conceivable way of constructing any form of motion picture presenting simultaneously all the changing features of even one region, even in outline. We are forced to distinguish between an historical and a geographical point of view, and in order to master the technique of either, we need to keep clearly in mind the distinction between the two.

3

Nature Reformation

Historical Background

Geographers are wont to boast of their subject as a very old one, extending, even as an organized science, far back to antiquity. But often when geographers in this country discuss the nature of their subject, whether in symposia or in published articles, one has the impression that geography was founded by a group of American scholars at the beginning of the twentieth century. Likewise many such discussions in the past have had only slight relation to similar discussions by European geographers, in spite of the well-known fact that geography has had a greater development in the universities of Germany alone than in those of this country.

European geographers, notably the Germans, have been much more generally interested in the study of the nature of geography than those in this country. Many of the questions raised here only in recent years have been very effectively discussed by certain German geographers since the beginning of the century. It seems desirable therefore to examine these discussions at Some length. If satisfactory answers can be found there to objections raised by our critics we may be spared the disturbance created by apparently new suggestions for radical departures from established lines of work, and those working in established lines may be assured of the value of their efforts. On the other hand if these many studies of the problem indicate that in certain respects we have been following false hopes, further loss of effort may be spared.

The studies of the nature of geography on which this paper is based are listed, together with illustrative studies, in the foregoing bibliography. Some effort was made to include in the former all of the more significant studies of recent years—though it is no doubt deficient in respect to the French literature, not to mention those in languages other than English, French and German. No such effort was made in connection with the illustrative literature however, the writer merely using whatever studies happened to come to his mind.

In a critical survey of this character, based almost entirely on the work of living colleagues in his field, a writer might prefer to omit specific references, but this easier choice is forbidden him because of his responsibility not only to the reader but also to those colleagues whose work has provided the material for the study. In particular, the critical position assumed places an imperative responsibility upon the writer to provide a means by which quotations, paraphrases, and critical comments may easily be checked against the writings concerned. Consequently a voluminous number of individual references are required. To avoid the disruption of the reader's thought created by so many footnotes, a simple reference system, suggested by the editor, is followed. The numbers in brackets provide the necessary basis for reference without, it is hoped, unduly interrupting the reading.

It would of course be absurd to suggest that American geographers were entirely unfamiliar with the concepts of geography developed in Europe. Arnold Guyot of Princeton, the first professor of geography in an American university, was a follower of Carl Ritter, but he had no successors. Far more important was the missionary work of Ellen C. Semple, whose presentation of Ratzel's concepts of anthropogeography set the course of human geography in this country for a quarter of a century. During the time period, William Morris Davis and Professor Albrecht Penck of the University of Berlin did more than any others before or since to establish mutual relations between geography in this country and Germany. Not only was there a close connection in their own work, but each taught for a year or so in the other country-Penck at Yale and Columbia, Davis at Berlin-and each published to some extent in the other language. Their field of mutual interest, however, was almost exclusively limited to physiography. The emphasis that Davis gave to this part of geography combined with Semple's exclusive consideration of human geography developed much the same form of dualism that was characteristic of an earlier period in Germany. Davis' attempt to bridge the gap was purely theoretical, as he himself made little effort to make the connections in practice.

For nearly two decades these two points of view dominated most of the thought of American geographers, both in their research and in their concepts of their subject as presented, particularly, in the presidential addresses of Brigham, Dryer, Barrows, and Whitbeck. The notable exception offered by Fenneman, influenced by the work of Hettner, appears to have made little impression until some years later. With that exception, geography was conceived of as a science of relationships between the natural environment and human activities. This "environmentalist concept" of geography was given its most careful statement by Barrows, who" reversing the usual form, speaks of geography, or "human ecology," as the study of man's adjustments to the natural environment.

A somewhat different point of view was introduced into this country from France by the publication in English translation of works of Vidal de la Blache and Jean Brunhes. Although both of these writers retained in theory the concept of geography as the study of relationships between man and the earth, in detail their work pointed in a different direction, and particularly tended to emphasize regional studies.

The greatest change in point of view, however, has been caused by the work of Carl Sauer in bringing the attention of American geographers to concepts developed in Germany nearly a generation ago. Through his three somewhat similar studies, published between 1925 and 1931, as well as by his marked personal influence, he has led many, particularly among the younger workers, to think of geography in terms of the study of material landscape features, both natural and cultural, and to consider these features according to their choreographic, or regional, interrelations.

American geography has therefore been markedly influenced by the work of European geographers, but it is notable that this influence has come through but very few contacts-for the most part through Davis, Semple, and Sauer. A few individuals, like Fenneman, or Wellington Jones, have been independently influenced, but have had little effect on the concepts of geography held by the rest. To be sure, many American geographers have followed the work of British geographers, but unfortunately the latter have been as provincial as ourselves if indeed not more so. The exceptions appear conspicuous in their lack of influence. Thus Chisholm, in his inaugural address at Edinburgh in 1908, abstracted at some length "the most comprehensive and to my mind the most illuminating recent series of papers on the scope of geography," namely Hettner's classic statement of 1905. But this view of the field, essentially, as we shall see, that of most German geographers, appears to be known to but few in England today. Similarly the views that Herbertson brought back from his studies in Germany made little impression in England until relatively recently. The great majority of English geographers appear completely unaware of the developments in German geography since Ratzel (note, for example, Bryan's frank admission of this fact in a footnote report of a conversation with one German geographer-whose name is misspelled and whose views are misstated. Most English geographers, depending largely on Semple for their knowledge of Ratzel, have tended greatly to overemphasize his importance in Germany and Dickinson and Howarth. In more recent discussions, however, Dickinson has strongly emphasized the concepts of German geographers 'of this century; further, a number of' English geographers have been influenced by the work of Passarge.

The situation in France has been somewhat similar. Vidal de la Blache and Vallaux gave great attention to Ratzel's work, but in the later methodological discussions of Vallaux and of Brunhes one finds little understanding of what German geographers since Ratzel have thought or achieved. Consequently the several publications in English of these French geographers have done nothing to lessen American ignorance of modern German geography.

In return, we may add, German geography has been but little affected by the methodological views developed in other countries. As exceptions, we may note the influence of the regional monographs produced under the direction of Vidal de la Blache, which undoubtedly helped to further the recent emphasis on regional geography in Germany; and Herbertson's efforts to establish "natural regions of the world," which for a time gained more attention in Germany than in England. The publication in German of lectures by Sir Francis Young husband has influenced a number of present German geographers, notably Banse and Volz. Apparently the only American geographer who has notably influenced German geography to date was W. M. Davis, whose developmental system of land forms classification was taken up by Penck, and is also represented in one of the standard German texts of geomorphology of which Davis and Braun were co-authors. German 'students of this subject continue to argue over the relative merits of the *"Davische"* system. On the more general aspects of methodology of geography as a whole, however, Davis appears to have had little influence in Germany.

These relatively limited associations in the' methodology of the subject are in marked contrast to the increasing degree to which American geographers are utilizing the research products of European geography. Koeppen has become almost a household word in our considerations of climatology; the work of Russian and German students has revolutionized our classification of soils; *Siedlungsgeographic* has' added new words like *Strassendorf* to our terminology; Passarge's concepts of *Landschaft* find expression in at least one recent American text; those interested in political geography have been made familiar with the developments in this field in Europe; and Joerg provided us with a survey of the whole field of European geography during the World War and post-War periods.

In spite of this development, however, the tradition apparently remains that any question as to the nature of the subject is a matter of personal opinion rather than one requiring study. Even the animated discussions that followed Sauer's challenging statements appear to have led few to examine the original studies on which his conclusions were based. With the exception of Hall's resume of the concept of regions and a short discussion by the writer, most geographers have been content with second-hand quotations, or have depended on their own consideration as to what geography ought to be. The most important exception is Finch's recent presidential address, published as this paper was nearly completed.

Reformation of the Nature

It has often been suggested that the question of the nature of geography is of minor importance-that what is needed is simply for geographers to work *"Es ist viel herumgedoktert worden an der Geographie"* Penck complains, and states that "progress in a science is not made by writing about methods, but by working methodically". But we may presume that Penck, who has published at least eight methodological discussions, does *not* mean that it doesn't matter whether we know where we are going so long as we are on the move. In a field in which the members who are officially recognized as "geographers" may, at a single meeting, read papers which to some of the hearers appear to belong in geology, climatology, soil science, economics, history, or political science, it is particularly necessary that we know what our field is, in order that we may hope to understand what its individual students are trying to do.

Even more important is the need for each individual who proposes to devote his professional life to the field of geography to have a clear picture of the scope and nature of that field. Not that any prescribed limits should be outlined within which everyone who is called a geographer must confine his work; no individual or group is given any authority in science to prescribe such limits nor to determine the character of the work to be carried on within them. But experience has all too clearly demonstrated that those who enter the field of geography with preconceived ideas of what it is or should be are afflicted with a feeling of dissatisfaction, if these ideas prove to be in conflict with what geography actually is. Whether this feeling of dissatisfaction be directed at themselves or at their colleagues, it represents an unhappy waste of intellectual energy deleterious to the effective progress of both the individual and the field as a whole. It is therefore one of the primary duties of those who attract young students into any particular field to give them so accurate a picture of the nature of that field that any later disillusionment will be unnecessary. But we cannot do this unless we ourselves know the nature of our field.

No one can read or listen to many of the discussions on the nature of geography by American geographers during recent decades without observing a marked motivation of reform. Evidently there is a certain general dissatisfaction with the field of geography, at least in its present state. To some degree at least these symptoms are echoes of similar symptoms that may be observed in the literature of European geographers, notably of the Germans.

Needless to say, a feeling of dissatisfaction with the present state of any field of knowledge—as distinct from a dissatisfaction with its essential nature-is a symptom of healthy vitality. But violent or arbitrary reactions to that feeling are not evidences of maturity. A bright-eyed youngster may observe errors in the way in which the pieces of a stained-glass window have been put together, and if he demonstrates his point by aiming stones accurately at the points in question he will create a notable disturbance but he will have done little for the improvement of the window. In many such cases the following statement of Humboldt seems appropriate: "It is far from my practice to find fault with endeavours, in which I have not myself made any efforts, because their success so far appears very doubtful".

In discussing one of the more violent reformers among German geographers, Cradmann has asked: "Is it really completely unavoidable that our science should be exposed naked before all the world, in that every new year a new 'reformer' arises who condemns the whole previous works to the foundations and feels himself called upon to erect something entirely new out of the ruins".

We have already suggested-and the later consideration of one or two of the more famous attempts to reform geography radically will demonstrate the suggestion-that a thorough dissatisfaction with the state of the field is a phenomenon whose explanation involves the critic as well as the object of his criticism. Although we are justified in keeping this possibility in mind there is no intention of dismissing in any such manner any proposals for reform in a field of study for which none of its students would claim perfection. Each of the important proposals of recent years will be analyzed on its own merits, in its relation to the historical development of geographic thought and in its relation to a logical concept of the field.

Several different tendencies may be observed in the various proposals that have been urged for the reform of geography. The current *Weltanschauung* in German thought finds expression both in the demand that geographers should give greater emphasis to the importance of individual men responsible for the historical development of the features of any region, and in the demand that geography must serve the national interest. To anyone familiar with the German geographical literature of the World War and post-War periods the suggestion that German geographers have been deficient in studying geography in Germany's interest is at least as surprising as to the elder German geographers themselves. On the assumption that this movement is not likely to make a positive impression in this country, we need not consider it further the interested reader will find pertinent discussions in Hettner, and Plewe.

A second tendency of greater importance in this country, seeks to add to geography some element that provides "problems." The "mere description" of the character of a region, even with the addition of interpretation, apparently, is not a problem, or at least not a problem in the sense in which that term is used in some other sciences. Consequently, it is thought necessary to add the element of change in time, to compare the situation of today with that of a previous time to study the effects of some particular change in one factor in a region upon the others, or, in general, to make geography "dynamic" rather than "static."

The remaining efforts to reform geography that need to be considered appear to be much more definitely motivated by a desire to make it what apparently it has not been, a proper science. In the first place, since science must deal with observable phenomena, geography, in order to be a science, should limit itself to observable phenomena, and these it finds in the visible "landscape" or, if it is claimed that some things are observable but not visible, geography must limit itself to material objects. These, in contrast with immaterial phenomena, we may observe and measure with some degree of accuracy and certainty.

Another suggestion is based on the assumption that each particular science has its own distinctive objects or phenomena to study. A science of geography must therefore somehow discover its distinctive objects of study. Most of the visible or material objects that it finds in areas have already been claimed, if not thoroughly exploited, as properties of other sciences. One possibility, therefore, is for the geographer to pick up whatever objects he can find that no one else has claimed, and mark these for his own.

Other geographers insist however that the proper objects of study for geography are the pieces of land which he calls regions. In order that these may be studied as other scientists study their objects, arranging them in classifications of types and developing scientific principles or laws about them, they must be concrete unitary objects. If one questions this assumption but accepts the major premise as to the nature of science, then the study of regions is not science but perhaps a form of art, which must be excluded from a "science of geography."

At this point one is reminded of a remark of Professor Whitehead when confronted by one of his students with a discontinuity in his logical reasoning: "What we need at this point is a beautiful phrase." The escape from the dilemma in regard to areas as objects is aided by one of those blessed words of double-meaning landscape." Since the landscape is the visible scene, including within it visible material objects, and may be limited in various undefined ways, it is called a unitary concrete object. But the landscape in its German form *"Landschaft,"* if not in English-is also the limited area of land, province or region. Hence the region or landscape is a concrete unitary object having form and structure and therefore subject to classification and the development of scientific principles concerning its relations.

The detailed examination of the nature of geography which this paper endeavours to present is not based on any assumption that geography is or ought to be a science-or that it ought to be anything other than it is. Assuming only that geography is some kind of knowledge concerned with the earth, we will endeavour to discover exactly what kind of knowledge it is. Whether science or an art, or in what particular sense a science or an art, or both, are questions which we must face free of any value concepts of titles. We will, therefore, not ask ourselves whether geography, as Douglas Johnson puts it, can "expect to enjoy the high prestige of the sciences," whether it "should pretend to a position of equality among them," or should accept a more humble though "honourable place along with history, economics, and sociology." Such questions naturally produce only the reaction that Johnson himself expressed: a refusal to "surrender geography's claim to a place of equality among the sciences until the scientific possibilities of the subject are more thoroughly explored than they have as yet been". Geography is thereby put on the defensive, ever attempting to be something that perhaps it cannot be.

On one aspect of this question we may clear the ground at the start. That geography could be "a science in the sense that geology and botany are sciences" is as impossible as it would be for botany to be a science in the sense that geology is' a science, or vice versa, or for either to be a science in the sense that physics and chemistry are sciences. Whatever geography is, its venerable if not honoured age would nullify the most enthusiastic efforts of any students to make it over into something entirely different. Geography is not an infant subject, born out of the womb of American geology a few decades ago, which each new generation of American students may change around at will. Granted that one must keep an open mind on any subject in science, we are hardly free, as one unnamed geographer claimed, to change our concept of geography each year as the spirit moves us.

In any established field, the problem of arriving at its proper definition is not so much a creative task, a problem in logical reasoning from *a priori* assumptions, as a research problem in knowledge itself. The philosopher, Kraft, in subjecting the field of geography to a critical examination to which we will refer in detail later-claims that the problem of the position of any subject within the totality of sciences is a work that extends beyond the particular scientific knowledge of the department concerned, is properly a problem "in scientific theory, *i.e.*, in the theory of cognition." But the function of such a research in the theory- of science cannot be to prescribe what geography should be, but rather to ascertain what it contains. Similarly Hettner wrote, in 1905: "The system of the sciences has been an historical growth; abstract designations of the sciences that tend to take no account of the historical development-unfortunately the methodological literature of geography has been particularly rich in such *a priori* conceptions-are foredoomed to unfruitfulness".

This point of view has recently been characterized by Leighly as blind adherence to "tradition" in the face of "strict logic". Even if the concept of geography that he defends could be shown to be logically sound a question to which we will come later-the argument is not to be won by devitalizing the historical evolution of a field of study as mere "tradition." When one consider the cultural origins of the inhabitants of the United States, one might claim that logically they should speak either a variety of European and African languages or a mixture in which each was represented in proportion to its part in the total population; it is merely a "tradition" of our schools that all should be taught something approximating "the King's English."

Growth of Knowledge

The growth of knowledge in any branch of science depends less on the particular scope for which it may find a logical justification, than on the continuity of its life, from the past through the present to the future. In the history of science-as in all history—common knowledge a Dong educated people so emphasizes the achievements of a few outstanding "masters" as to lead one to suppose that each of them, starting from scratch, had single-handed accomplished his wonders. In geography, we have a striking example in Alexander v. Humboldt. He is frequently presented as a pioneer without predecessors, a lone figure who explored America and returned to establish the science of physical geography. In reality, as we shall shortly observe, it may be questioned whether Humboldt would have become a geographer at all, had it not been for the younger Forster, and the latter's geographic work was very largely dependent on the elder Forster; his antecedents we do not know. Furthermore Humboldt throughout his work depended on a veritable host of other workers, both of earlier times and of his own time, as the multitudinous references in his writings indicate.

New Suggestions

If, in contrast, each new generation of geographers, or each individual geographer presents new suggestions, untested either by example or by study of the previous development of the subject, there is introduced, as Hettner observed, a feeling of unrest and insecurity among geographers that handicaps the successful cultivation of the field and, in addition, causes workers ill other sciences to become skeptical of the whole subject. American geographers are all too familiar with such results. If we are to continue to experience violent shifts of the helm-formerly toward physiography, then toward environmentalism, now toward landscape studies, tomorrow to "the topography of art" and thereafter who knows whither—our ship will beat around with ever-changing aim, hence aimlessly, and will arrive nowhere.

Throughout the preparation of this study the writer has assumed that few if any readers are interested in the particular view that he as a single geographer may have of the field. The privilege of presenting statements of that kind is reserved, one presumes, for the masters of our field whom this Association elects to its presidency. The writer's concern on the contrary, is to present geography as other geographers see it-or have seen it in the past. Since the geographers of any particular time and country represent but a small part of the whole field, it will be necessary to extend our examination both into other countries and into the past. If we wish keep on the track-or return to the proper track, as was suggested in a rennet symposium we must first look back of us to see in what direction that track has led, our first task will be to learn what geography has been in its historical development.

Reversing the usual procedure, therefore, we will not start with any particular concept of the nature of science, but rather with the field of geography as it has been produced. Having ascertained its essential characteristics we will endeavour to find what logical position it holds among the different branches of knowledge. The first of three major portions of the paper is therefore concerned with the question "What is geography?"

On the foundation thus established we will be in a position to examine various suggestions that have been made for improving geography. This examination of major problems in geography today' constitutes approximately half of the total study. The reader will understand that the brief statement of suggestions given above does not do justice, even in outline, to the ideas of the students proposing them. Because many of these students have made distinguished contributions to geographical research their theoretical suggestions for its improvement demand our careful attention. On the other hand we may presume that they themselves will wish that their ideas should not to be taken on authority but should be subjected to critical examination. Since in many cases the significance of a theoretical suggestion can most clearly be seen where it has been applied in practice, our examination may also require a critical survey of the material products. In both cases the reader is particularly requested to keep in mind that it is only the validity of the ideas of the nature of geography that concerns us here. The informed reader will recognize from the bibliography that I have confined myself largely to the writings of geographers of established distinction. Since the standing of these writers is in nearly every case based on their non-methodological works, it cannot be affected by any discussions in this paper. It is the methodological ideas of these writers that we are concerned to question: are these, either in terms of logic or of fad, valid statements of the nature of geography or acceptable suggestions for improvement within geography? If these questions cannot be' answered in the positive, it will *be* necessary to consider a sound logical basis on which geography may develop as a broad but unified field.

Finally, by way of conclusion, we may seek to relate the field of geography, in its essential characteristics, to the general nature of knowledge, to determine, that is, what kind of a field of study geography is. Whether it is properly to be called a science or not, we will leave to those more interested in terminology; if we know definitely what kind of a study it is, we need not concern ourselves over titles.

The writer wishes particularly to stress, that in considering the suggestions of his colleagues, his purpose is not to engage in dialectic arguments, seizing any opportunity that lack of clarity of statements, might offer, but rather to understand the intention of the suggestions made. In particular he has no wish to introduce into American geography a form of personal argumentation that has unhappily marked' the methodological discussions of German geographers in recent times. With Obst, his purpose, is to seek for the maximum degree of understanding and agreement possible, not to establish any particular direction in, geography but to make possible a common basis for geographic work, so that the geographic work itself-the only real purpose-may be furthered to the highest degree.

Historical Survey

In a brief examination of published studies on the history of geography, John K. Wright commented on the lack of any adequate treatment of this subject in English. We have numerous excellent treatments of the history of American geography, notably those by W. M. Davis, Douglas Johnson, and Colby. Attention may also be called to the recent publication in German literature of two separate surveys of recent developments in American geography, by Broek, of the University of California, arid by Pfeifer, who has also worked in the same institution; but note Platt's critical comment. The development of geography in this country, however, can only be understood in the light of earlier and contemporary developments in Europe; for a thorough treatment of this subject we still look in vain in literature. Since Wright first noted this lack, in 1925, Dickinson and Howarth have published a useful survey of the field, but one that unfortunately depends largely on foreign interpretations of the work in Germany, and gives but the briefest sketch of the developments in that country since Ratzel.

There can be no question that the foundation of geography as a modern science was primarily the work of German ,students. In whatever country one starts, the study of the development of geography leads backward to the work of Humboldt and Ritter in the early part of the nineteenth century. Further, though in many countries geographic work has become more or less independent, major changes in geographic thought are found to be expressions frequently very belated, of the developments in German geography. It is particularly unfortunate therefore that we have in English hardly even an outline of the development of geographic' thought in Germany during the past century, so that nearly all American geographers have been trained in complete ignorance of the methodological background of their subject. German students, in contrast, have readily available not only the original writings of the period, but also many historical studies of the development of geographic thought.

To present this history adequately would require a book in itself. For our present purpose we must confine ourselves to the consideration of the major characteristics of geography as a field of study at each of several successive stages in its development. Our concern is not with the development of geographic work, nor with the relative importance of different writers or groups in the production of geographic material, but rather with the establishment and changes in thought concerning the nature of geography, as represented however in geographic works as well as in direct methodological studies.

Our examination of earlier work in geography has a two-fold purpose. We have already emphasized its importance in providing us with an understanding of the character of the field in which we work. The second purpose will be to observe what earlier writers have concluded in regard to problems which are of current concern in the methodology of geography, for there are few of these that have not claimed the attention of geographers in the past-in most cases, repeatedly in the past. As Braun observed, "there will be no improvement in the methodological confusion as long as we geographers continue the habit of completely ignoring the writings of our predecessors in this problem. . . and attempt to construct a building with stones that the neighbor has perhaps long ago recognized as useless or insufficient for a better arrangement". "Any progress in methodology," wrote Obst, "must be built on the past. No methodological change therefore without knowledge and consideration of past understandings and concepts".

These two purposes of our historical study cannot be attained if one enters the study with a pre-established concept of what geography should be, and on these terms searches among the writers of the past for those whose ideas are in agreement with ours, whom one then places in opposition to those who appear to have taken a different view of geography. In any methodological debate, to be sure, one may look for supporting arguments wherever he may find them-whether in the present or the past-but such presentation is not to be confused with an historical study which aims at in objective presentation of the past; on the contrary it leads easily to a distortion of the past development of geography and of the ideas of earlier geographers in order to fit the purposes of the particular debate. We will therefore attempt to consider the concepts of previous geographers, so far as is humanly possible, independent of our own particular views.

4

Methodological Discussion

Chorographic Science

Common-Sense Justification : The question is frequently raised whether the study of the areal differentiation of the earth's surface provides a sufficient function for geography, whether a "science" of geography does not need to seek more vital problems. Stated in these terms the question raises the problem of the nature of science in general, the discussion of which is postponed to the final section of this paper. *For* the present we will merely assume that whatever has engaged the intellectual curiosity of peoples of all times-including many of the most eminent thinkers-is a subject worthy of advanced study. Geography as a chorographic (or chorologic) study has always found its justification in the widespread desire of many people to know what other parts of the world are like, just as history finds ample justification in the common desire to know what happened and what things were like in past times. Further, since the ordinary person actually knows but partially the area in which he lives, just as he understands but little of what is happening in the world in his lifetime, "home geography" is as necessary as current history.

"The purpose and the irreplacable significance of geography," says Volz, "is that it teaches us to know the space we live in, the surface of the earth". We do not need to prove that the study of the world in which we live is worthy of serious minds. "Geography assumes the responsibility for the study of areas," as Sauer says, "because there exists a common curiosity about that subject. The fact that every school child knows that geography provides information about different countries is enough to establish the validity of such a definition. No other subject has preempted the study of area. . . .' If one were to establish a different discipline under the name of geography, the interest in the study of areas would not be destroyed thereby. The subject existed long before the name was coined.... The universality and persistence or the chorologic interest and the priority of claim which geography has to this field are the evidences on which the case for the popular definition may rest".

Character of Study

The character of the study that is presented to geography in the mandate from common intellectual curiosity is unquestionably a highway complicated one; from the point of view of the specialized sciences, each dealing with a restricted class of phenomena, it may well appear bewildering in its scope and character, in its heterogeneity of phenomena. But this mandate permits of no avoidance of the problem; to attempt to break the field up into separate, specialized parts is not to create various other sciences but to eliminate geography, For man, as Vidal observed, is Interested above all In the totality of conditions in any area. "From the beginning of the era of voyages and explorations it is the spectacle of the social diversities associated with the diversities of places, that has aroused his attention." Consequently, "for most of the ancient writers from whom geography received its titles of origin, the concept of 'country' (contree) is inseparable from its inhabitants; the: particular character of a country is expressed no less by the means of nourishment and the physical aspect of its human population than by the mountains, deserts, and rivers that form their surroundings".

It should not be implied from the very general character of our statement of the function of geography-to learn what the different parts of the world are like—that we are to be satisfied with answers of equally general nature. On the contrary, to provide complete, detailed, accurate, sound, explanatory, and organized answers to such general questions will tax the very best abilities of the mature student.

Obviously there are many different ways of studying the world, but since men as individuals, and as individual groups, do not in any very full sense live in the entire world, but each in a relatively restricted area of the world, one of the most significant methods of studying the world is to study it by areas. Geography, in brief, can demand serious attention if it strives to provide complete, accurate, and organized knowledge to satisfy man's curiosity about how things differ in the different parts of the world, just as history in similar fashion strives to satisfy his curiosity about what things were like in the past; and just as history considers the past in terms of periods because men live and things happen together only within a limited space of time, so geography must consider the world in terms of limited areas within which things are closely associated.

Few, therefore, will question the value of geography in the education of man, by which I mean not merely in the schools, but rather in the full sense of widening of one's perception of the world. There exists, however, a remarkable but not uncommon misconception of the relation of education and research according to which a subject may be valuable enough for instructional purposes, but not be worthy of research by mature scholars. To be sure, the mere reorganization of information already known into an outline suitable for teaching purposes is a pedagogical, rather than a research task—though not necessarily unworthy of nature scholars. If, however, by regional, geography were meant simply the arrangement of geographical data by regions, it is doubtful if a course of such character would be worthy to be taught at any level. The thorough regional study which aims at an adequate understanding of the spatial relations and interrelations of phenomena found together within any region is no mere problem of organization, but requires research calling for all the intellectual ability of mature scholars. If it is granted that the results are worthy to be taught to others, how can there be any question that the problem is worthy of research? For, while we cannot ask our ordinary students to learn all that has been learned by research, we can hardly justify asking them to learn anything that was not, at some time, worthy of research by mature scholars. All that we dare to teach, at any level, elementary as well as university, requires research original research to ascertain if it be true, and repeated research to check its truth.

Another question often raised, though the answer might seem obvious, is whether geographers are the appropriate *persons* to secure and supply the knowledge of areas. Certainly it is true that, just as history is studied and written by many who are not by training historians, there are many non-geographers willing and able to supply knowledge of areas. Ralph Brown's recent study of the geographical material available for the Atlantic Seaboard of the United States in the period around 1800 shows that the world is not dependent alone on professional geographers for such information. If these be lacking, or fail to perform their central purpose, others may, in some fashion, take their place. But his study likewise illustrates the vast amount of work necessary simply to test the dependability of material gathered by untrained travellers, naturalists, agricultural students, literary writers or statesmen, and, particularly to organize it effectively into regional study.

From its beginning, geography has had to depend on the effective descriptions of travellers and literary writers as well as upon the more scientific, quantitatively measured and logically related studies of trained students. The development of the subject, Hettner notes, has brought these two divisions together and simple description has been replaced by an interpretative description which looks for causal connections.

Nevertheless the regional geographer must still assume a task which is in some ways similar- to that of the artistic (or literary) describer-that of presenting a synthetic description of an area. Undoubtedly he will envy the superior ability of the artist to give us the feeling of an area-I know of no better description of the feeling of the Spanish meseta than that of Washington Irving, in the introduction to *The Alhambra.* But before a geographer hands over his task to the artistic worker he would do well to consult artists and writers, or students of art and literature, to be sure that they will accept his assignment. The subjective impression which the artistic worker (including both painters and literary writers) receives from a landscape or region, and which he desires to convey to others, is something very different from the objective description which the geographer must attempt to provide. So far as pure description is concerned the method of the geographer is photographic in character, with the distinctive personal reactions of the observer reduced to a minimum, a limitation which the artist cannot accept. For this purpose the geographer has techniques that are unknown to the artistic worker. To a large extent he may depend on cartographical presentation. Further, by explaining the interrelations between the phenomena which he describes he should be able to give the reader a much fuller picture of the character of the area he is describing than any non-geographer can hope to do, for its character consists not' merely of what is seen at first sight, but includes the complex of relationships of the things seen.

Finally, however effective may be the descriptions of artistic writers in presenting the character of an area, these descriptions cannot be expected to satisfy scientific standards of knowledge. One does not expect the writings of a Walter Scott or a Francis Parkman to supplant methodical historical narratives. Similarly, only trained geographers can provide an objective, quantitatively measured, scientifically interpretative, and dependable presentation of an area. If, at the same time, this requires a high degree of skill in the use of maps, diagrams, and the written word, geographical literature is not lacking in such skillful presentations. That is a question of style, and style in the presentation of knowledge does not remove it from the sphere of science. But if geography is to remain always objective, as Penck insists, the describer is not to express his feelings, but to impart how he has comprehended the things objectively.

Relation with other Sciences

Although geography in view of its long history can be expected to remain essentially that which it has been, if it is to have a productive future its nature must bear logical analysis. In particular any clarification or limitation of its nature will depend on the place which reason assigns to it in some logical arrangement of the divisions of science—"science" that is, we repeat, in the broadest sense of organized, objective knowledge.

Although relatively few of the students who have attempted to determine the field of geography from purely logical considerations have given this fundamental problem adequate attention, geography nevertheless stands on no weak grounds. As we have noted in our historical survey, our field received for many years the attention of one of the great masters of logical thought. In the introduction to his lectures on physical geography, Immanuel Kant presented an outline of the division of scientific knowledge in which the position of geography is made logically clear. The point of view there developed has proved so satisfactory, to others as well as to this writer, both in leading to an understanding of the nature of geography and in providing answers to all questions that have been raised, that it seems worth while to quote at some length from Kant's original statements.

"We may classify our empirical knowledge in either of two ways: either according to conceptions, or according to time and space in which they are actually found. The classification of perceptions according to concepts is the logical classification, that according to time and space the physical classification. Through the former we obtain a system of nature, such as that of Linnaeus, through the latter, a geographical description of nature.

"Thus, if I say that cattle are to be included under the class of quadrupeds, and under the group of this class that have cloven hooves, this is a classification that I make in my head-that is, a logical classification. The system of nature is likewise a register of the whole: there I place each thing in its class, even though they are to be found in different, widely separated places.

"According to the physical classification, on the other hand, things are regarded according to the places that include them on the earth. . . . In contrast to the system of nature with its division by classes, the geographical description of nature shows the places in which things are actually to be found on the earth. For example, the lizard and the crocodile are fundamentally the same animal: the crocodile is merely an extremely large lizard. But they exist in different places: the crocodile in the Nile, the lizard on the land. In general, we consider here the scene *(Schauplatz)* of nature; the earth itself, and the regions where the things actually are found, in contrast with the system of nature which is concerned with the similarity of forms. . . . .

"Description according to time is history, that according to space is geography."... "History differs from geography only in the consideration of time and area *(Raum)*. The former is a report of phenomena that follow one another *(nacheinander)* and has reference to time. The latter is a report of phenomena beside each other *(nebeneinander)* in space. History is a narrative, geography a description. . . .

"Geography and history fill up the entire circumference of our perceptions: geography that of space, history that of time."

In much the same manner Alexander v. Humboldt described the position of geography among the sciences. As early as 1793, nearly a decade before the publication of Kant's lectures, he had distinguished clearly between the sciences that study nature according to categories of objects, those that consider all the phenomena of nature according to changes in time *(Naturgeschichte)*, and those that consider the objects as interrelated together in space *(Geognosie* or, later-following *Kant-Weltbeschreibung*, including descriptive astronomy and *Erdbeschreibung* or geography). There was a greater similarity between the two latter groups, he noted, than between either of them and the systematic sciences. The student of Humboldt's works finds little difficulty in correlating all of his studies to this schematic division of science, which, we may add, he restated at various times in his life-finally, and in most detail, in the first volume of the *Kosmos*.

Ritter likewise, we saw, evidently conceived of the relation of geography to history and to the systematic sciences in much the same manner, though he does not appear to have stated the comparison as clearly as either Kant or Humboldt. During the subsequent period of shifting views in German geography-the drift toward historical studies-in the school of Ritter, and the counter movement that Peschel introduced in the opposite direction-the logical position of geography as defined by Kant and Humboldt was lost sight of and not restored until the early part of this century. Likewise in the spread of geography to other countries, Kant's views were of little significance, and ho' 'ever important may have been the influence of Humboldt whose works were translated into many languages—his views in regard to the position of geography' among the sciences appears to have been overlooked.

Logical Basis of their Concept : Methodological Discussions

The methodological discussions in Germany in last quarter of the century-notably the Leipzig address of Richthofen-stimulated a number of younger German geographers to consider the logical basis of their concept of the field. In the ensuing discussions almost every German geographer of standing has taken part, either in brief statements associated with research publications or more frequently in separate methodological articles. In contrast with the purely personal expositions of each individual geographer's concept of the field that characterizes so many of the methodological addresses of English and American geographers, a large number of the studies of German students of this problem are based on scholarly consideration of previous treatises and of the course of historical development of geographic thought. Although this contrast may be, due in part to a difference in scholarly attitudes, it is no doubt also a result of the easy access that German students have to the records of past geographic thought, since those have been written largely in their language.

There are however many indications in the more recent literature that the standard of scholarship set by the generation of German geographers now retiring from the academic stage-many of them before reaching the age of retirement is regarded by many of their successors as outmoded. From some of these students one hears demands for "freedom" for new ideas to replace concepts of the past that are summarily rejected. On examination these demands prove to rest in large part on misunderstanding or ignorance of what earlier writers actually thought. Although this change has been commented on by many German geographers it should not be supposed that there are not numerous exceptions in both groups. The former group included some who, as Braun noted, were wont "'to pass sovereign over previous works" and there are many among the younger writers whose methodological discussions demonstrate a high degree of scholarship. We will not prejudge the individual students for the reader: in the later sections of this paper ample evidence will be offered for any Judgements he may care to make.

Whatever one may think of the methodological studies of the German geographers just mentioned, the intensity and thoroughness with which they have examined the problems concerned make it seem hardly conceivable that a serious student would wish to express conclusions of his own without first examining their efforts. If he has done so and finds those to be in error, we have the right to expect him to explain his reasons for so finding.

Of all the German geographers of recent times-indeed of all geographers—of any time—none has worked so long and so continuously on this problem as Alfred Hettner. His predominant importance in establishing a major degree of unity of thought among German geographers of our time has recently been testified to by one who could not be regarded as a follower, Albrecht Penck. As one of the first to have chosen geography as a life-work from the beginning, Hettner was spared the uncertainty and shifting of attitude that affected those—like Penck himself-who had transferred from one field of science to another. (Hettner, we may add from a personal communication, received his doctor's degree under Gerland but "was always in opposition to him in regard to the conception of geography, which he had taken over from his first teacher, Kirchoff," and later modified under the influence of Richthofen, under whom he subsequently studied. But Hettner, Penck continues, is, no mere theoretical methodologist; his work began with scientific fieldwork in the Andes-in the footsteps of Humboldt-and, in spite of a physical injury that appeared to incapacitate him for further field work, he has, adds Penck, "with iron strength maintained his personal touch with the objects of geography." On the other hand Hettner's considerations of geography have not been based on a view limited to that field alone; he has the whole of science in view and; as a critic from the field of philosophy, Graf, observed, is "the only one among the methodologists of geography in recent times who had found the connection with philosophy". Finally, the clear logic of Hettner's discussions, the clarity of his expression, and the consistency of his thinking, have enabled him to establish a clear and unified concept' of' the field of geography in the long series of methodological treatises that Solch rightly calls "classics".

American Geography

American geographers are, of course, by no means ignorant of the importance of Hettner's position in German geography. Unfortunately, however, the most influential interpretation of the concepts of modern German geographers in this country, by Sauer, pays little more attention than praise to his writings; in so far as his views are considered they are misinterpreted. In considerable part the other views that are presented are those that Hettner, and with him the majority of his colleagues, have rejected. While these statements will require demonstration in later sections, it is appropriate to note here characteristic expressions of opinion of German geographers. Solch describes Hettner as "without doubt the leading methodologist and one of the most important leaders generally in German geographical science," whose work is "an imperishable gain for the more sure foundation and general recognition of geography". Similar opinions have been expressed by Penck, as already noted, Philippson, and Tiessen. Krebs evidently regards the failure to consider "even" Hettner's essays in the discussion of methodological problems as damning in itself.

The number of German geographers who base their methodological discussions on Hettner's writings is far too numerous to list in full, but in addition to his own students we may note the following: Gradmann. It is also significant that certain students who in recent years have been objecting to the *Weltanschauung* of the older group, direct their attack particularly at Hettner. While not hesitating to come to the defence, Hettner, in proper objective fashion states: "My own importance in the construction of the methodology has been exaggerated; likewise I do not believe that my conception is 'completely peculiar.' I believe only that I have clearly expressed and methodologically established what was actually present in the development." Note also his ironic comment on another critic: *"Spricht er doch sogar einmal geschmackvoll von einer hettnnerisierung der Geographie"*.

Outside of Germany, Hettner's first major essay of 1905 made a notable impression on Chisholm, and more recently Dickinson and Howarth have recognized his importance: "While Vidal de la Blache established the method of regional description on a local basis, and Herbertson, the concept of major natural regions on a world basis, the main credit for interrelating the two and analyzing exhaustively the method of treatment lies with the modern school of German geographers . .. led by the veteran pioneer Hettner . . .". In Japan, Inouye and others have recognized the major importance of Hettner's work for geographic methodology. Apparently only in France and this country is it supposed that discussions of the nature of our subject can ignore his work.

In contrast with many other students, Hettner's methodological writings were closely related to his actual geographic work. As noted above, he had made regional field studies in South America before he wrote on methodological questions. Likewise, it was not in the study, but on the shore of Lake Titicaca that he outlined his view of general, or systematic, geography. Discussed in various subsequent methodological essays this finally took actual form nearly forty years later in the four-volume set entitled "*Vergleichende Landerkunde*". Long before that, his concepts of regional geography had been presented not only in theory but in the two volumes of his *Landerkunde,* in which, as Solch reports (in his review of the more recent edition) "he has applied the principles previously developed and thereby provided the test of example—indeed many of his theoretical problems had their seed in the *Landerkunde*".

Our concern here is with Hettner's methodological studies which form the major foundation for the subsequent discussions throughout this paper. They were preceded by the brief informal statement of the field of geography with which Hettner introduced the *Geographische Zeitschrift,* which he founded in 1895. The brief survey of the development of geography, published in 1898, provided the historical introduction for the series of methodological essays that began in 1903 and appeared intermittently throughout his forty years of editorship. The fundamental study is that of 1905 on "*Das Wesen und die Methoden der Geographie*", the greater part of which will still repay careful study.

Most of the essays published before 1927 were collected, in part rewritten, in the volume that Hettner described as his principal life-work—*Die Geographie, Die Geschichte, ihre Wesen und ihre Methoden*. Inouye, in referring to this book, in an article in the *Geographische Zeitschrift,* as "the most valuable work in the history of geographical methodology," may have been expressing Japanese courtesy; nonetheless he was stating the obvious. It would be of great value of geography in English-speaking countries if Hettner's volume, together with selections from his subsequent discussions, could be published in English translation. The lengthy citations in this paper will present, it is hoped, a more adequate picture of his philosophy of geography than has hitherto been available in English. Any who wish to take issue with him, however, should study carefully his complete treatment in Die Geographie and his later statements; the writer has repeatedly found that ideas presumed to be original had already been discussed in those writings.

In considering the logical position of, geography among the sciences, Hettner, like both Kant and Humboldt before him, proceeds not from the consideration of particular branches of science but from a view of the whole field of objective knowledge. His major division of the field of science is essentially the same as that of his illustrious predecessors, although he was not made aware of Kant's analysis until after he had written his own, and apparently did not then or later realize the close similarity to Humboldt's views.

If, as is quite possible, each of these students arrived at essentially the same concept 'independently of each other, one may have all the more confidence in its validity. In any case, Hettner has developed the concept most fully and, since the greater part of our subsequent discussion of the nature of geography depends on a clear understanding of the comparison of our field to that of other branches of knowledge, it will be well to present his statement at length from the most recent complete statement, that of 1927.

"Reality is simultaneously a three-dimensional space, which we must examine from three different points of view in order to comprehend the whole; examination from but one of these points of view alone is one-sided and does not exhaust the whole. From one point of view we see the relations of similar things, from the second the development in time, from the third the arrangement and division in space. Reality as a whole cannot be encompassed entirely in the systematic sciences, sciences defined by the objects they study, as many students still think. Other writers have effectively based the justification for historical sciences on the necessity of a special conception of development in time. But this leaves science still two-dimensional, we do not perceive it completely unless we also consider it from the third point of view, the division and arrangement in space.

Systematic Science : The systematic sciences ignore the temporal and spatial relationships and find their unity in the objective likeness or similarity of the subjects with which they are concerned. The common distinction between the natural and the social sciences is such a systematic distinction. Within the, natural sciences we have, first, the sciences of the minerals and stones, mineralogy and petrography, of the plants, botany, of the animals, zoology, and with them, but *for* other reasons developed as a separate science, paleontology, the science of fossil plants and stones. Only later was the study of the earth body and its circles of phenomena taken up by separate disciplines. Systematic social sciences are these of language, of religion, of the state (political science), of economics, etc.

"For the historical sciences the material relations are objects of only incidental importance. Rather, in their study they unite a number of things of entirely different systematic categories and' obtain their unity through the point of view of the temporal progression of events in time. If these followed each other merely by chance, and the course of various groups of phenomena were independent of each other, science could be content with merely the systematic studies. But the connection of different times, which we express by the word development, and the connection in the same time, make necessary a separate historical study. The studies of the development of a single kind of phenomena which, for that kind alone, takes account of both the points of view named, such as the history of the animal world or the history of art or of literature or constitutional history, occupy an intermediate position between the systematic and the historical sciences.

"With the same justification as the development in time, the arrangement of things in space demands special study. Together with the systematic, or material, and the chronological or historical or time-sciences, there must develop chorological or space sciences.

"There must be two of these the one has to do with the arrangement of things in the universe; that is astronomy. . . the other is the science of the spatial arrangement on the earth, or, since we do not know the interior of the earth, we can say on the surface of the earth."

The logically close relation of astronomy and geography, which was especially characteristic of Humboldt's cosmos, may on first thought seem confusing. That is because astronomy is often incorrectly expressed simply as an applied mechanics, or on the other hand as the science which studies the heavenly bodies as its concrete objects. On the contrary it has been a major function of astronomy to discover planets and vast numbers of stars, both by the application of mechanics to the arrangement and relations of the heavenly bodies and by exhaustive search of the heavens. The investigation of the nature of the individual bodies is a part, but only a part, of its field.

The comparison between astronomy and geography, to which we will return in a later connection, is discussed in detail by Hettner, who notes in particular that a large part of so-called mathematical or astronomical geography is properly a part of astronomy. The logical similarity of the two fields is also noted by W. M. Davis and Vallaux. One important difference however should be noted here: in heavenly space there are clearly defined and separate bodies which can be accepted without question as unit objects of systematic study, whereas in the space formed by the earth's surface there is no such separation into parts that can be accepted as naively given objects-not even the apparent divisions of the lands by the seas correspond in total character to the divisions of the heavenly bodies by empty space.

Geography, then, as the second chorological science is the study of the spatial arrangement on the surface of the earth. "If no causal relations existed between the different places on the earth, and if the different phenomena at one and the same place on the earth were independent of each other, no special chorological conception would be needed; since, however, such relationships do exist, which, by the systematic and historical sciences, are comprehended only incidentally or not at all, we need a special chorological science of the earth or the earth surfaces.

"The historical and geographical viewpoints require that time or space be placed in the foreground and form the unifying bond of the scientific study. As history studies the character of different times, so geography studies the character of different areas and places—'the earth-spaces filled with earthly contents,' to use Ritter' expression—it studies the continents, regions, districts, and localities as such".

View of few Geographers : Few geographers evidently have felt themselves sufficiently grounded in philosophy to discuss Hettner's system of the sciences. The somewhat confused discussion by Schluter will be considered in a later connection. Leutenegger, in an erudite, rather abstruse philosophical study, concludes that, though Hettner has not provided a satisfactory system of the sciences neither has anyone else; but Hefiner's at least is logical and geography fits logically into it. Among others who have attempted to examinee the subject from a somewhat different point of view, but arrive at conclusions that are in major part in agreement with those of Hettner may be mentioned Michotte.

Unfortunately, philosophers interested in the problem of the division of the sciences seldom have a knowledge of the actual field of geography sufficiently adequate to enable them to judge the system that Hettner, in agreement with Kant and Humboldt, has outlined. An objection that Wundt raised has but little importance for geography. Hettner spoke in terms of abstract (mathematics), abstract-concrete (physics and chemistry) and concrete sciences (zoology, economics, etc.), but it seems dear that this distinction can be eliminated as I have done in quoting-without affecting Hettner's major thesis; if anything it strengthens it. Wundt's own suggestions for the position of geography among the sciences are evidently based on a very inadequate knowledge of the field; it is doubtful if any geographer would accept his conclusions. Benjamin's discussion of geography (as reported by Finch-A. C. Benjamin, An Introduction to the Philosophy of Science, New York appears to be based on the concept of geography as the study of relationships. If his conclusions could be justified on that basis, Finch has shown that they do not apply to geography conceived as a chorographical science.

Two German students of philosophy who have also a considerable command of at least the methodology of geography have recently considered the nature of our field. Graf's treatment is ,based categorically on a particular philosophical theory and, though we will find much of value in his critical discussion, his attempt to fit a field of science to a particular philosophical theory does not prove satisfactory for the science. It is characteristic of his method of reasoning that he dismissed Hettner's system solely on the basis of the objection that Wundt had made-apparently assuming that the discovery of a philosophical error in a line of reasoning necessarily dismisses all subsequent conclusions, without demonstrating that the point in question is in any way necessary to the reasoning.

In a brief but very instructive examination of the unity of geography as a science, Kraft does not specifically discuss Hettner's system of the sciences, but accepts at least its major conclusion for geography: "Stones, plants, animals and man, in themselves objects of their own sciences, constitute objects in the sphere of geography insofar as they are of importance for, or characteristic of, the nature' of the earth surface." This gives a clear answer to the question of how far man may be included in a unified geography: "it is the reciprocal relations of man and the earth-area *(Erdraum)* that determine and limit the treatment of man in geography".

In his further consideration, Kraft appears to overlook a major element in Hettner's definition of geography. It is not merely the earth surface as a whole that forms the object of geographic study, but rather-as the comparison with history emphasizes-the differences in the different areas of the earth surface. Consequently, meteorology, though a study confined to the earth surface (in the sense in which we use that term) is not a part of geography, but is a systematic science basically necessary for climatology which does form a part of geography. Likewise this viewpoint provides 'at least a partial answer to the question for which Kraft found no logical answer, namely, granted that geography includes the study of man, how can it justify logically the high position which it inevitably gives to human, cultural phenomena. While human phenomena contribute perhaps but a minor part in comparison with nature to the formation of the earth surface considered as a whole, in the consideration of its differential character, they constitute a factor of major importance—for example, in the comparison of Java and Borneo.

Geography and History

The purpose of translating so much of these discussions is to emphasize the value of the comparison of geography and history. If the geographer, especially the regional geographer, finds any difficulty in developing or maintaining his concept of his field, the suggestion is that he apply this comparison as a test. If the field in which he has for years been working is suddenly and vigorously attacked, so that he fears the whole development is threatened with collapse, let him reread the attack, substituting for the word "geography" the word "history," for the word -"region" the word "period." The result will appear to be the devastation of the field of history-there cannot be, he will learn, a science of history.

The history of different periods is simply a form of narrative in which incommensurable phenomena must be presented in integrated form; therefore it could be far better developed by literary writers than by professional historians. The historian then, in the sense of the general student of the history of a period, in contrast to specialists in topical history, presumably should give place to literary writers. It is hardly necessary to demonstrate that the conclusion is absurd; the necessity and validity of history as a technical subject for trained scholars is sufficiently established.

It is clear, however, that the history of a period, like the geography of a region, is a very different sort of study from the systematic sciences. The latter, whether in the natural or the social fields, select like things wherever they find them, whenever they have been found, compare and study them in themselves and in the relationships which, in relatively simple manner, they bear to each other. Both the historian and the geographer must study a far greater range of different kinds of facts. Kroeber suggests that history is an endeavour at descriptive integration of phenomena in contrast to the method of the systematic sciences which is "essentially a procedure of analyses, of dissolving phenomena in order to convert them into process formulations". The same might well be said of regional geography, with the difference that history is concerned with integration with reference to time, geography with reference to space.

When geographers concerned with the scientific standing of their subject think of "science" only in terms of such systematic fields as geology, zoology, or economics, they have reason to doubt whether regional geography can possibly hope to attain recognition as a "science". A "science of regions" in this sense may well be impossible, not merely as an ideal, but even as a working goal. But if geography is to look for its comparisons to history-and, essentially, to history alone—then the objections fall to the ground. Provided one has no emotional concern for the term "science," one may suppose that geographers will not heed to be discouraged if their subject can occupy a place in the world of knowledge comparable with that of history.

If this comparison is so important, one may well ask why it has not been more generally recognized. Certainly the similarity has not been entirely overlooked. Barrows made use of the comparison in his presidential address, without following logically the direction in which it leads. Sauer also suggests the comparison in one of two places, but makes relatively little use of it. Possibly he was diverted from following Hettner in this regard because he confused Hettner's concept with that which "sees in the 'Where' the distinctive quality 'of geography," in contrast with history, concerned with the 'When'. Consequently, those who have depended solely on Sauer for their knowledge of German methodology, have not been made aware of the importance of the similarity.

A few however have utilized it effectively. W. M. Davis made use of the comparison in defending geography against the charge of being merely a composite subject. Whittlesey has made a suggestive comparison of the educative value of geography and history; and the English geographer Crowe has made good use of the comparison which he appears to have learned from Fairgrieve; [Fairgrieve gives no references] .

Scientific Character

That most American geographers, when considering the scientific character of their subject, tend to look for comparisons with the systematic sciences may well be a natural result of the particular development of geography in this country. Long and close association with one of the systematic sciences, geology, tended to direct our comparative thought far from history, and indeed from the systematic social sciences, toward the systematic natural sciences. While this undoubtedly had great value in developing standards of scientific method, notably in measurement and accurate mapping, it did not contribute a proper understanding of the essential nature of our field. Further, because of that bad ground, it is not surprising that the greater part of the productive work of American geography has not been directed on complete geographic studies of areas, but rather has consisted of systematic studies of particular features found in different areas. Wheather these are uplifted peneplains, stream deposits, temperature efficiency, potato fields, steel mills, or political boundaries they represent areal features arbitrarily removed from their geographical settings, for purposes of systematic study. Necessary as such work in "systematic geography" ("general geography" among Europeans) may be, overemphasis upon it tends to direct our attention away from the specific nature of the subject. Almost all geographers agree that the clearest view of geography is to be seen in regional geography *(Landerkunde)*.

The comparison with history, then, emphasizes for us the special character' of geography as an integrating science cutting a cross-section through the systematic sciences. This figure of Hettner's, which could be derived also from Kant, differs significantly from Fenneman's well-known diagram of the relation of geography to other fields. Geography does not *border* on the systematic sciences, overlapping them in common parts on a common plane, but is on a transverse plane cutting through them. A diagram in solid geometry would therefore be required to illustrate this.

Character of Geography

The lengthy discussion of the comparison of geography with other sciences is not, as one might suppose, concerned merely with a question of suitable or unsuitable analogies, but rather with fundamental differences in point of view which have led to very important differences both in the concepts of the subject and in the manner of treatment of specific problems. Since the beginning of the century there has been fairly general agreement among German geographers on the chorographical nature of geography as the study of areas, and this concept has found its way into various other countries, as we have seen, and, in particular, has been accepted by a large number of the active research workers in this country. Few will question that this general agreement made possible a more cooperative, and hence more effective, development of geography. But among these students who agree in general, there is radical disagreement on three important considerations, disagreements arising in large part at least from the manner in which the different students compare and relate geography to other sciences. These disagreements are concerned with

(1) the place of historical development in geographic work,

(2) the criteria for selection of phenomena in geography, and

(3) the fundamental concept of the unit areas which the geographers are to study, each of these questions will be considered in detail in following sections.

5

The Interruptions

Evolutionary Theory

Cyclic concept was probably postulated first in geology by Scottish geologist James Hutton in 1785 when he propounded the concept of 'cyclic nature of the earth history' and the dictum of 'no vestige of a beginning; no prospect of an end' and 'present is key to the past'. The concept of cyclic nature of the earth history was later on transformed into the concept of 'uniformitarianism' which states that 'the same physical processes and laws that operate today, operated throughout geologic time, although not necessarily always with the same intensity as now." Probably based on this inheritance of Hutton's concept of the cyclic nature of the earth history and the evolutionary concepts of Charles Darwin (origin of species through natural selection) William Morris Davis (1850-1934 A.D.) presented his concept of cycle of erosion under the title of 'geographical cycle of erosion' in the year 1899 though the concept of 'complete cycle of river life' was already propounded by him in 1889.

The Davis' model of geographical cycle of erosion was based on the basic concept of 'sequential change in landforms through time like the evolution of an organic life'. According to Davis landforms undergo sequential changes through time (passing through three stages of youth, mature and old). The earth's surface is affected by two types of forces viz. (i) endogenetic forces and (ii) exogenetic forces wherein endogenetic forces create vertical irregularities on the earth's surface by forming several types of relief features of different dimensions whereas exogenetic processes originating from the atmosphere (rivers, wind, glaciers, sea waves, groundwater, periglacial processes etc.) try to remove the vertical irregularities created by the endogenetic forces and ultimately become successful in bringing down the reliefs to low featureless plain called as a peneplain. The whole period of the creation of relief features by endogenetic processes and their destruction by exogenetic processes is called cycle of erosion which Davis defined as follows, "Geographical cycle is the period of time during which an uplifted landmass undergoes its transformation by the process of land sculpture ending into a low featureless plain-a peneplane".

The concept of 'geographical cycle' of Davis was severely criticised by German geographers and the term 'cycle' was described as confusing and hence untenable. Walther Penck though accepted the basic concept of cycle of erosion but rejected the Davisian model of geographical cycle and propounded his own model of cycle of erosion. Inspite of severe criticism in Germany the Davisian model of geographical cycle of erosion was adopted by most of the contemporary and subsequent geomorphologists all over the world. It may be safely argued that Davisian model of cycle of erosion, say the first general theory of landform development, dominated the entire field of geomorphology and geomorphological investigations right from its inception in 1899 to 1950 throughout the world. His 'cycle of erosion' was basically concerned with the evolution of landforms in humid temperate areas but the cyclic concept was later on applied to almost all the geomorphic processes e.g., arid cycle of erosion (W.M. Davis, 1905), glacial cycle of erosion (Davis, 1900, 1906), marine cycle of erosion (Davis, 1912 and D.W. Johnson, 1919), karst cycle of erosion (Beede, 1911 and Cvijic, 1918), periglacial cycle of erosion (L.C. Peltier, 1950) etc.

Modifications of the Theory

C.H. Crickmay suggested modifications in Davisian model of 'geographical cycle' in 1933 and described the process of peneplanation to be more powerful and effective than Davis' process of peneplanation in the evolution of landforms. According to Crickmay the end product of the cycle of erosion would be panplain and not the peneplain. L.C. King proposed a new cycle of erosion named as 'the cycle of pediplanation' to explain the characteristics and evolution of landforms of arid and savanna regions of Africa as he found Davisian model of geographical cycle unfit to explain the landforms of the aforesaid regions. J.C. Pugh (1966) and M.F. Thomas (1966) propounded the concept of 'savanna cycle of erosion' to account for the development of landforms of semi - arid savanna regions of Africa. A.N. Strahler (1950), J.T. Hack (1960) and R.J. Chorley (1962) rejected the evolutionary concept of landform development as advanced by W.M. Davis and his followers and pleaded for the concept of 'time-independent landforms' instead of Davisian concept of 'time-dependent landforms' and advanced the concept of 'dynamic equilibrium model' of land form development.

Recently, 'tectonogeomorphic model' of Marie Morisawa (1975, 1980) 'episodic erosion theory' of S.A. Schumm and R.W. Lichy (1965) etc., have been suggested to explain the landform development. These models are, in fact, modified forms of Davisian model of landform development.

Davis' Geographical Cycle : William Morris Davis, an American geomorphologist, was the first geomorphologist to present a general theory of landform development. In fact, his theory is the outcome of a set of theories and models presented by him from time to time e.g. (i) 'complete cycle of river life', propounded in his essay on "The Rivers and Valleys of Pennsylvania' in 1889, (ii) 'geographical cycle' in 1899, (iii) 'slope evolution' etc. He postulated the cyclic concept of progressive development of erosional stream valleys under the concept of 'complete cycle of river-life', while through 'geographical cycle' he described the sequential development of landforms through time.

The general theory of landform development of Davis is not the 'geographical cycle' as many of the geomorphologists believe. His theory may be expressed as follows :

> "There are sequential changes in landforms through time (passing through youth, mature and old stages) and these sequential changes are directed towards well defined end product-development of peneplain."

The basic goal of Davisian model of geographical cycle and general theory of landform development was to provide basis for a systematic description and genetic classification of landforms. The reference system of Davisian general theory of landform development is "that landforms change in an orderly manner as processes operate through time such that under uniform external environmental conditions an orderly sequence of landform develops" (R.C. Palmquist, 1975). Various models were developed on the basis of this reference system e.g., normal cycle of erosion, arid cycle of erosion, glacial cycle of erosion, marine cycle of erosion etc. Thus, 'geographical cycle' is one of the several possible models based on Davis' reference system of landform development.

Davis postulated his concept of 'geographical cycle' popularly known as 'cycle of erosion' in 1899 to present a genetic classification and systematic description of landforms. His 'geographical cycle' has been defined in the following manner. "Geographical cycle is a period of time during which an uplifted landmass undergoes its transformation by the process of land sculpture ending into low featureless plain or peneplain (Davis called peneplane)."

According to Davis three factors viz. structure, process and time play important roles in the origin and development of landforms of a particular place. These three factors are called as 'Trio of Davis' and his concept is expressed as follows" Landscape is a function of structure, process and time" (also called as stages by Davis' followers). Structure means lithological (rock types) and structural characteristics (folding, faulting, joints etc.) of rocks. Time was not only used in temporal context by Davis but it was also used as a process itself leading to an inevitable progression of change of landforms. Process means the agents of denudation including both, weathering and erosion (running water in the case of geographical cycle).

Model of Penck : It may be pointed out that German scientist Walther Penck pleaded for the rejection of Davisian model of geographical cycle based on time-dependent series of landform development and presented his own model of 'morphological system' or 'morphological analysis' for the explanation of landscape development. The main goal of Penck's model of morphological system was to find out the mode of development and causes of crustal movement on the basis of exogenetic processes and morphological characteristics. The reference system of Penck's model is that the characteristics of landforms of a given region are related to the tectonic activity of that region. The landforms, thus, reflect the ratio between the intensity of endogenetic processes (i.e. rate of upliftment) and the magnitude of displacement of materials by exogenetic processes (the rate of erosion and removal of materials).

Penck is perhaps the most misunderstood geomorphologist of the world. It is not yet sure whether he used the word 'cycle' or not in his model of landform development. Penck's views could not be known in true sense and could not be interpreted in right perspective because of (i) his incomplete work due to his untimely death, (ii) his obscure composition in difficult German language, (iii) ill-defined terminology, (iv) misleading review by W M. Davis and (v) some contradictory ideas.

According to Penck landform development should be interpreted by means of ratios between diastrophic processes (endogenetic,-, or rate of uplift) and erosional processes (exogenetic, or rate of vertical incision).

Penck is supposed to have deliberately avoided the use of stage concept in his model of landscape development either to undermine the cyclic concept of W.M. Davis or to present a new model. According to O.D. Von Engeln (1960) "Penck found escape from the concept of cyclic change marked by the stages youth, maturity and old age". In the place of ,stage' he used the term entwickelung meaning thereby 'development'. Thus, in the place of youth, mature and old stages he used the terms aufsteigende entwickelung (waxing or accelerated rate of development), gleichformige entwickelung (uniform rate of development) and absteigende entwickelung (waning or decelerating rate of development).

Normal Cycle of Erosion

The cycle of erosion by fluvial processes (running waters or rivers) is called normal cycle of erosion because of the fact that fluvial processes are most widespread (covering most parts of the globe) and most significant geomorphic agent. Even water also plays important role in glacial, and arid regions. W. M. Davis considered humid temperate areas as the most normal case for fluvial cycle of erosion but this claim is debatable.

The normal cycle of erosion begins with the upliftment of any landmass with reference to sea level. As the land rises, the rivers are originated and their erosional work starts. The rate of uplift in the beginning far exceeds the rate of erosion with the result absolute relief (absolute altitude from sea level) and relative relief register increase. After some time upliftment of the land stops and erosion becomes more active. The land area, tectonically, remains stable i.e., there is crustal stability for long period of time during which there is neither upliftment nor subsidence of land area. There is progressive development of river valleys in sequential order and the whole land area progressively passes through three successive stages of youth, mature and old (senile or penultimate) and is ultimately transformed into low featureless plain of undulating surface. Thus, the Penultimate end product of normal cycle of erosion is called peneplain which is characterized by undulating surface with residual convexo-concave low hills known as 'monanocks', unakas' and 'mosores'.

Thus, the land area has to pass through the successive stages of its development right from the upliftment of landmass to its transformation into peneplain of exceedingly low reliefs. W.M. Davis has divided the whole duration of normal cycle of erosion into three successive stages of youth (juvenile), mature (equilibrium) and old (penultimate or senile) and each stage has been further divided into three substanges e.g., early, middle and late (for example, early youth, middle youth and late youth and so on). Thus, the landscapes also become young, mature and old with the advancement of normal cycle of erosion. Like landscape development through three successive stages, the development of river valleys also passes through three successive stages of their development and the rivers become young or youthful rivers, mature rivers and old rivers. The following are the characteristic features of successive stages of the normal cycle or erosion.

Youthful Stage : Consequent streams (which follow the regional slope) are originated with the upliftment of land area due to endogenetic forces. In the beginning, the streams are less in number and short in length. Very few tributaries of the master consequent streams are originated. The slopes are dominated by numerous rills and gullies rather than big streams. These rills and gullies lengthen their longitudinal profiles (increase their lengths) through headward erosion. Gradually and gradually the main streams deepen their valleys. The origin and evolution of tributaries of master streams give birth to the development of dendritic drainage pattern. The rivers are continuously engaged in rapid rate of downcutting of their valleys (valley incision) because the transporting capacity of the rivers is maximum due to high velocity of flow rate and kinetic energy because of very steep channel gradient. High transporting capacity enables the rivers to carry big boulders (tools of erosion) of fairly good size (large size) and calibre (angular boulders) which help in the pothole drilling of the river beds. It may be mentioned that pothole drilling is the most active and powerful process of vertical erosion (valley deepening) in the juvenile stage of the normal cycle of erosion.

The valley becomes very narrow and deep with almost vertical side walls due to continuous active downcutting of the valley floors at exceedingly fast rate. The valley side slopes are convex in plan. Thus, the resultant juvenile valleys are V-shaped and are called gorges and canyons. The valley floors are studded with numerous pot holes which are the result of pothole drilling. The interstream areas or water divides (land area between the valleys of two major streams) are extensive and wide and these are least affected by denudational processes because valley widening by lateral erosion is less effective in the early and middle youth stages. The valley thalwegs (longitudinal profiles of the rivers) are characterized by numerous rapids and waterfalls which always recede upstream. Most of the waterfalls and knick points disappear by late youth. The rivers are underloaded (not having the required amount of sediment load according to their transporting capacity) and thus available energy is more than the work to be done. The rivers are well integrated by the end of youth when maximum relative reliefs are formed.

River capture is the most characteristic feature of the juvenile stage of the normal cycle of erosion. Main rivers having steeper channel gradients and more volume of water capture smaller streams of relatively low channel gradient through headward erosion.

Mature Stage : Marked valley deepening through vertical erosion during youthful stage results in pronounced decrease in channel gradient and consequent decrease in flow velocity with the result the arrival of early maturity is heralded by marked decrease in valley deepening due to (i) decrease in channel gradient, (ii) decrease in the velocity of river flow, (iii) decrease in the transporting capacity etc. Consequently, valley widening through active lateral erosion dominates over valley incision through downcutting. The convex slope of valley sides is progressively transformed into uniform or rectilinear slope and the gorges and canyons characterized by deep and narrow valleys are replaced by broad and flat valleys.

The rivers deposit big boulders at the foothill zones due to sudden decrease in channel gradient and hence marked decrease in the transporting capacity of the rivers. These materials form alluvial fans and alluvial cones. The gradual expansion of these fans and cones due to their continuous growth result in the formation of extensive piedmont plains through the coalescence of several fans and cones. Interstream areas or water divides are continuously narrowed due to backwasting caused by active lateral erosion and valley widening. Thus, interstream areas are transformed into narrow ridges. The major river erodes down to its base level (sea level) and becomes 'graded'. Thus, the longitudinal profile of the master river becomes the profile of equilibrium wherein there is balance between available energy and the work to be done i.e., balance between the transporting capacity and total sediment load to be transported. Because of marked decrease in channel gradient rivers adopt sinuous courses and develop numerous meanders and loops in their courses. Extensive flood plains are formed due to sedimentation of alluvia. Rivers frequently change their courses because of gentle to level slopes of the flood plains. Numerous ox-bow lakes are formed due to straightening of highly meandering loops. Deposition of sediments on either side of the river valleys leads to the formation of natural levees.

Old Stage : The old stage is characterized by further decrease in channel gradient, almost total absence of valley deepening, decrease in the number of tributary streams and flattening of valleys. Tributary streams also attain, the base level of erosion and are graded. Lateral erosion and consequent backwasting eliminates most of innerstream areas. Valleys become broad and flat characterized by concave slopes of valley sides. Downcutting of the valleys is totally absent. Weathering processes are most active. Thus, lateral erosion, downwasting and weathering continuously degrade the land resulting into gradual lowering of absolute altitude and water divides.

Interstream areas and water divides are remarkably reduced in height and are changed to lowland but they still rise above the surrounding areas. Transporting capacity of the rivers becomes minimum because of very low channel gradient and thus the rivers become overloaded. Consequently, sedimentation becomes most active during this stage. The rivers adopt highly meandering courses. The extensive flood plains with level to gentle slopes (2°-5°) and very low channel gradient make the river flow so sluggish that the main channel of the river is divided into numerous distributaries and thus the river becomes braided. Valley sides are bordered by extensive natural levees which are also known as bluffs which denote the farhtest limit of recurrent floods of the concerned rivers. Rivers deposit and form extensive deltas at their mouths if other environmental conditions remain favourable for delta formation.

The entire landscape is converted into extensive flat plain of undulating surface except a few residual convexo-concave hills which project above the general flat surface and thus break the monotony of relief less flat plain, called as peneplain. These residual hills, the result of differential erosion, are called monadnocks on the basis of Monadnock hills of the North-East Applachians in New England region (USA). The whole landscape is dominated by concave slope, minimum available energy, both potential (because of very low height) and kinetic energy (due to very low channel gradient) and maximum entropy (means maximum disorder of relief, as the whole area is characterized by featureless peneplain).

This ideal normal cycle of erosion can pass through its all the three stages (i.e. youth, mature and, old) and the peneplain can be formed only when the region remains in stand still position for longer duration of crustal stability but the availability of such condition is of remote possibility because the earth is very much unstable. The plate tectonics have also revealed that plates are always mobile and hence long period of crustal stability is not possible. Thus, the smooth functioning and completion of normal cycle of erosion is very often disturbed due to tectonic events and also due to climatic changes. The disturbance in the cycle of erosion is called interruption of normal cycle of erosion which is caused by changes in sea level (either due to upliftment or subsidence of oceanic bottoms or due to glaciation and deglaciation during great ice ages, like Carboniferous Glaciation of the Gondwanaland during Carboniferous period and Pleistocene Glaciation of the northern hemisphere during Pleistocene period), upliftment or subsidence of land areas, volcanic eruptions and climatic changes. The interruption of normal cycle of erosion by volcanic eruptions or climatic changes is called 'accident'. The interruptions caused by negative changes in base level (lowering of base level-maximum limit of vertical erosion by rivers) due to fall in sea level (because of the subsidence of the oceanic floors) and upliftment of landmass are called rejuvenation, which means renewed vigour of erosive capacity of the rivers. Rejuvenated rivers are again engaged in valley deepening at much faster rate and thus the cycle of erosion is driven back to youthful juvenile) stage. Several interesting landforms, like valley in valley topography or multi-storeyed valleys, paired terraces, incised meanders, nick points and nickpoint water falls, uplifted peneplains, deeply entrenched gorges and canyons within broad flat valleys etc., are formed due to rejuvenation. The landforms resulting from several cycles of erosion, whether complete or incomplete, are called polyeyelic landforms or multicyclic landforms, e.g. Chhotanagpur region represents multicyclic landforms.

Interruptions and Polycyclic Beliefs

Meaning : Davisian cyclic concept envisages gradual sequential changes in uplifted landmass through the stages of youth, mature and old culminating into the development of low featureless plain called as peneplain. The completion of such generalized and ideal form of cycle of erosion depends on tectonic stability of longer period of time, which is seldom possible in the nature as the earth is unstable. Thus, the cycle of erosion is liable to frequent interruptions which cause inequilibrium condition in the cyclic model. Any sort of obstacle in the normal functioning of cycle of erosion is called interruption of cycle. The basic causes of interruption may be climatic or tectonic or both. Tectonic factors are related to positive and negative movement of base levels of erosion. The interruptions in cycle of erosion caused by positive movement of base level (due to subsidence of landmass or rise in sea level) shortens the cyclic time as it advances forward e.g. if the cycle is in youth stage, it may advance (due to positive movement of base level) to mature stage or if it is in the mature stage, it may advance to old stage. Conversely, negative change in base level of erosion (due to upliftment of landmass or fall in sea level) lengthens the cyclic time as the cycle is pushed backward e.g., if the cycle is running through late mature stage and there is either sudden upliftment of landmass or fall in sea level, the cycle goes back to youth stage characterized by renewed vigour of valley incision known as rejuvenation. Thus, the cycles punctuated by interruptions are called as interrupted cycles which lead to occurrences of several cycles in a region: Such cycles are called polycycles. If the polycycles (multi-cycles) occur in succession, they are called successive cycles of erosion and the landscapes resulting therefrom are called polyclic or multi-cyclic landscapes. Applachian Regions of the U.S.A. and Chhotanagpur Region of India present typical examples of polyclic landscapes where several cycles have been completed. Paired terraces, valley in valley topography, and incised meanders of the Damodar Valley at Rajroppa in Hazaribagh (Jharkhand) and valley in valley topography and paired terraces, Dhunwadhar falls and incised meander of the Narmada river at Bheraghat near Jabalpur (M.P.) are indicative of rejuvenation and polycyclic reliefs.

It may be pointed out that the interruption of cycle of erosion is generally divided into two categories i.e., if the interruption is such that the period of cycle is either lengthened (due to rejuvenation) or shortened, it is simply called as interruption of cycles. It may also he called as base level change interruption. If the interruption is such that the chapter of ongoing cycle is closed and a new cycle is initiated after a gap of long time, such interruption is called as accident which is caused either due to climatic changes or volcanic eruptions. Suppose, a fluvial cycle of erosion is running through mature stage, and if there is a widespread volcanic fissure flow leading to pouring of immense volume of lava resulting in the obliteration of all the existing streams, then the chapter of cycle of erosion would be closed. New set of streams would appear on new surface after the lava is cooled and solidified and a fresh cycle of erosion would start.

Vulcanicity and Climatic Changes : The chapter of any cycle of erosion is closed due to permanent interruptions caused by volcanic eruption and climatic changes. The widespread volcanic fissure flow causes upwelling and pouring of immense volume of basaltic lava which covers larger area and obliterates surface drainage and reliefs which results in the permanent interruption and closure of the existing chapter of existing fluvial cycle of erosion. The fresh cycle of erosion may start only when the fissure flow ceases, lavas are cooled and solidified, new surface is formed and new sets of streams are initiated. It may be pointed out that such interruptions in the fluvial cycle occurred over Indian peninsula during Cretaceous period when Deccan lava flows covered vast areas of the Deccan plateau including Chhotanagpur plateau and even southern parts of the Vindhyan upland causing closure of the Jurassic cycle of erosion. The new Tertiary cycle was initiated only after the Deccan trappean lavas were cooled and solidified and monsoon climate set in.

Climatic interruptions (accidents) occur due to major changes in the climate of the concerned region, for example, if the fluvial cycle of erosion in a humid region is passing through mature stage, and there is sudden climatic change leading to onset of either extreme dry conditions or extreme cold conditions, then the cycle of erosion is interrupted to such an extent that the chapter of current cycle of erosion is closed and another set of cycle either arid cycle (if the climate becomes extreme hot and arid) or glacial or periglacial cycle (if the climate becomes glacial/ periglacial or sub-glacial) sets in. It may be mentioned that if there are minor changes in the climate, then the chapter is not closed, rather it (cycle) is either augmented or slowed down. Suppose, if the climate becomes more humid leading to increased amount of rainfall, then the surface runoff and stream discharge will automatically increase which would cause local interruption in the cycle by accelerating the rate of erosion (case of rejuvenation), the effects of which may spread over larger areas.

Base level Changes : Any change in the base level of erosion (determined by the sea-level) causes interruptions in the cycle of erosion or may initiate new cycle whether the previous cycle is complete or not. Changes in base level of erosion are basically of two types i.e. positive change (due to rise in sea-level) and negative change (due to fall in sea-level). The base level changes caused due to sea-level changes are eustatic (widespread). Base level of erosion also changes due to tectonic factors i.e. subsidence of landmass causes positive change whereas upliftment of landmass is responsible for negative change. It may be mentioned that such type of subsidence or upliftment of landmass may be or may not be related to sea-level. Positive change of base level (or rise in base level) shortens the period of cycle of erosion whereas negative change of base level (or fall in base level) lengthens the cyclic time. Secondly, positive change of base level of erosion is indicative of accelerated alleviation (deposition) whereas negative change causes accelerated erosion (rejuvenation). It is evident that rejuvenation is the most important factor of interruption in the fluvial cycle of erosion and genesis of polycyclic (multi-cyclic) reliefs and hence it needs detailed discussion.

Rejuvenation : Rejuvenation simply means acceleration of erosive power of the fluvial process (rivers) caused by a variety of factors. Rejuvenation lengthens the period of cycle of erosion. For example, if the cycle of erosion is passing through senile stage (old stage) characterized by gentle channel gradient, sluggish river flow and broad and shallow alluvial valleys, after rejuvenation (caused either due to substantial fall in sea-level or due to upliftment of landmass) the cycle is interrupted and is driven back to juvenile (youth) stage characterized by steep channel gradient and accelerated valley incision. Rejuvenation is of 3 types —

1. Dynamic Rejuvenation Causes
   (i) upliftment in the landmass

(ii) tilting of land area

(iii) lowering of outlet

2. Eustatic Rejuvenation Causes : changes in sea-level due to
   (i) diastrophic events (subsidence of sea-floor or rise of coastal land)
   (ii) glaciation causing fall in sea-level
3. Static Rejuvenation
   (i) decrease in the river load
   (ii) increase in the volume of water and consequent scream discharge due to increased rainfall or melt-water
   (iii) increase in water volume of the main river due to river capture

The Causes

As pointed out earlier, the basic cause for rejuvenation of fluvial cycle of erosion (acceleration of erosive power of the river) is negative change in the base level of erosion (which is determined by the sea-level) which is caused by a host of factors. Negative change of base level of erosion is always related to negative change in sea-level (fall in sea-level) which is also called as eustatic movement as it is widespread and global phenomenon. It may be pointed out that fall in sea-level (and hence causing negative movement/change in base level of erosion) steepens the channel gradient resulting into increased kinetic energy of the fluvial process which resorts to valley incision with renewed vigour.

The eustatic negative change in sea-level is caused during glacial ages when most of seawater is locked on the continents as thick cover of ice sheets. The consequent lowering of sea-level causes steepening of channel gradient of streams which are, in fact, rejuvenated and are engaged in active downcutting of their valleys. The Pleistocene glaciation of the northern parts of N. America and Eurasia caused widespread rejuvenation in the temperate and tropical zones. Four river terraces of the Red river of the U.S.A have been related to four periods of advancement of ice sheets i.e., Nebraskan, Kansan, Illinoin and Wisconcin glacial periods of the Pleistocene Ice Age.

Negative change in sea-level causing rejuvenation locally and regionally is also caused because of subsidence of sea-floor in relation to coastal land due to tectonic factors.

Local or regional upliftment of landmass causes interruption in the fluvial cycle of erosion and rejuvenates the fluvial processes (streams). Such type of regional rejuvenation caused by secular rise in the landmass has been reported from several parts of the Chhotanagpur Highlands of Bihar which experienced 3 phases of upliftment in response to three episodes of upliftment of the Himalayas during Tertiary period. The Patlands of the Ranchi plateau and Palamau uplands (Jharkhand) were subjected to an upliftment of 305 m resulting in the interruption of fluvial cycle of erosion and rejuvenation of the N. Koel river and its tributaries. The nick points and resultant waterfalls on Burhariver (Burhagaugh falls 142m, Gutamghaugh falls, 36.57 m and Gharaghughra falls, 7.62 m)-a tributary of the N. Koel river, on Pandra river (Ghagri falls of 43 m), on the Sankh river (Sadnighaugh falls of 61 m), on the Jori river (Jalmighaugh falls of 37 m), on the Ghaghra river (Nindighaugh falls of 45.72 m) etc., indicate rejuvenation.

Lowering of outlets of streams also causes rejuvenation due to release of extra volume of water in the concerned river. Such rejuvenation (due to increase in the volume of water) also occurs when the water supply suddenly increases due to river capture (supply of extra water of the captured stream to the captor stream).

Topographic Discordance

The typical landforms resulting from interruptions in the fluvial cycle of erosion and from rejuvenation resulting in the formation of a mosaic of poly or multi-cyclic landforms include topographic discordance, valley in valley or multistoreyed valleys, uplifted peneplains, incised meanders, paired terraces, nick points etc.

Topographic discordance refers to the creation of older topographic forms above and younger forms below. In other words, when the topographic accordance or uniformity from the top of the river valley to its bottom is not maintained rather is disturbed due to interruption in fluvial cycle of erosion caused by rejuvenation, the resultant topographic expression is called topographic unconformity or discordance wherein the upper part of the valley reveals the sign of senile or mature stage whereas the lowest part of the valley belongs to youth stage. The river develops flat and shallow valley at the end of mature and beginning of senile stage but if there is sudden negative change in the base level of erosion caused either by fall in sea-level or upliftment of landmass, the river is rejuvenated and begins active downcutting of its valley due to increased erosive power. Thus, a deep and narrow valley is formed within flat and broad valley. This new deep and narrow valley is flanked by terraces on its either side which represent earlier older valley. Such topography is called 'valley in valley topography' or 'two storeyed valley' o r'two-cycle valley'. By the march of time the rejuvenated river deepens its valley to the new base level and thu, forms second broad and flat valley within the first broad and flat valley formed during first cycle of erosion. Suppose, if there is again upliftment of landmass or fall in sea-level, the cycle is again interrupted and the river is rejuvenated. Consequently, again a narrow and deep valley is formed within previously formed two storeyed valley. Now, there are two paired terraces on either side of new deep and narrow valley. Thus, the resultant form of the valley is called three storeyed valley. Such topographic forms, where there are more than on valley in a single river's cross profile, are called 'multi-storeyed or multi-cyclic valleys'.

Examples : The Damodar valley at Rajroppa in Hazaribagh is a typical example of polycyclic valley or topographic discordance which is characterized by two-storeyed valley. The Damodar river developed its broad and flat valley of senile stage before the onset of Tertiary upliftment. The river was rejuvenated due to upliftment of landmass during Tertiary period caused by the side effects of the Himalayan orogeny and thus the Damodar excavated its new deep and narrow valley of youthful stage within its broad and flat valley of senile stage. The Bhera river coming from over the Ranchi plateau makes a water fall while joining the Damodar river and thus presents an example of a hanging valley. Such topographic discordance is also observable in the alley of the Narmada river at Bheraghat (downstream from Dhunwadhar falls, 15 k.m. away from Jabalpur city in M.P.).

Formation of Uplifted Peneplains

Three storeyed valleys have been located in the Himalayas where the rivers were rejuvenated thrice due to three major episodes of upliftment of the Himalayas during Tertiary orogeny. Nearly all the major rivers of the U.P. Himalayas are flanked by three sets of terraces on their either side.

Uplifted peneplains are formed due to interruption caused by rejuvenation consequent upon regional upliftment. The uplifted peneplains are represented by their remnants of accordant summit levels which rise above the general ground surface of the present-day planation surface. Uplifted peneplains are, in fact, the results of successive cycles of erosion wherein several fluvial cycles of erosion are completed in succession. Three uplifted peneplains have been identified in the Applachians (which are indicative of successive phases of upliftment, consequent rejuvenation and cycles of erosion) e.g. (from older to younger) viz. Schooley peneplain (after Schooley Mountain), Harrisburg peneplain (after Harrisburg Mountain), and Sammerville peneplain. The Patland of the Ranchi Plateau is a typical example of uplifted peneplain which is higher than the central Ranchi plateau (610 m a.m.s.l). The granitic-gneissic surface (910m a.m.s.l) of the western highlands has a capping of 154 m thick basaltic lava (now weathered to laterites) cover of Cretaceous period. Prior to Cretaceous lava flow the entire Ranchi plateau, including the present western highlands, was peneplained by Jurassic period, the western part of which received lava cover of 154 m thickness during Cretaceous period. This western part (610 m + 154 m lava) was uplifted by 305 m in Tertiary epoch and thus the granitic- gneissic surface of 915 m height lying below 154 m thick lateritic basalt is an example of uplifted peneplain. The North Koel and its numerous tributaries have dissected patlands and segmented them into several small tableaus locally known as 'pats' (which are fine examples of mesas and buttes) such as Netarhat pat, Khamar pat, Rudni pat, Jamira pat, Raldami pat, Bangru pat etc.

Incised meanders — Five alternative terms are in use for river meanders viz. incised meanders, entrenched meanders, intrenched meanders, in closed meanders and ingrown meanders. Incised meanders are the representative features of rejuvenation and polycyclic reliefs and are developed through vertical erosion leading to valley incision consequent upon renewed erosive power due to rejuvenation. The narrow and deep meanders formed due to accelerated rate of valley incision caused by rejuvenation within simple broad meanders (having wide and shallow valleys) developed by lateral erosion during 1st cycle of erosion are called incised meanders which are further divided into (i) entrenched meanders (having uniform slopes of both the valley sides of meander loops), and (ii) ingrown meanders which have unequal slopes of valley sides wherein one side of the valley representing concave side is deeply undercut and the outer side (convex side) is characterized by gentle valley slopes. The meandering valley of the Karo river downstream from Pheruaghaugh falls at the southern margin of the Ranchi plateau has been considerably incised due to rejuvenation and hence presents an ideal example of incised meander. The Damodar gorge near Rajroppa is typical example of incised meander. Similarly, Bheraghat gorge of the Narmada near Jabalpur (M.P.) is fine example of incised meander.

Knick Point

Knick point, simply called as nick point or only 'nick' represents breaks in slope in the longitudinal profile a river caused by rejuvenation. This is why nick point is also called as head of rejuvenation which registeres gradual recession upstream. These breaks in channel gradient or nick points denote sudden drops of elevation in the longitudinal profile of the rivers and allow water to fall down vertically giving birth to waterfalls of varying dimensions. These are called as nick point falls or simply hickfalls. Hundru falls (76.67 m) on the Subarnarekha river (near Ranchi city), Johna or Gautamdhara falls at the confluence of the Raru and Gunga rivers (to the east of Ranchi city), Dassam falls (39.62 m and 15.24 m) on the Kanchi river (east of Ranchi city), Burhaghaugh falls (148 m) on the Burha river, a tributary of the North Koel river, Dhunwadhar falls on the Narmada river (near Jabalpur, M.P.), major falls of Rewa plateau-M.P. (e.g. Chachai falls - 127 m on the Bihar river, Kevti falls - 98 m on the Mahana nadi, Tons or Purwa falls - 75 m on the Tons river, Odda falls - 145 m on the Odda nadi etc.) are the examples of nick points caused by rejuvenation.

Paired terraces are also significant features of poly-cyclic reliefs. Three pairs of terraces are found in the Himalayas indicating three phases of uplift and consequent rejuvenation.

6

Significance of Weathering

Nature and Concept

The process of disintegration and decomposition of rocks in situ is generally called weathering. It means weathering is a static process. According to C.D. Ollier (1969) "weathering is the breakdown and alteration of minerals near the earth's surface to products that are more in equilibrium with newly imposed physico-chemical conditions." According to P. Reiche (1950) "weathering is the response of minerals which were in equilibrium within the lithosphere to conditions at or near its contact with the atmosphere, the hydrosphere, and perhaps still more importantly, the biosphere." It may be pointed out that rocks are never in permanent equilibrium rather they are in equilibrium only momentarily and thus W.D. Keller (1957) has pleaded for the deletion of 'which were in equilibrium' from Reiche's above definition of weathering. B.B. Polynov (1937) has very precisely defined weathering as "the change of rocks from the massive to the clastic state."

Arthur Holmes has presented more elaborate definition of weathering which also includes the processes of weathering. According to him "weathering is the total effect of all the various subaerial processes that cooperate in bringing about the decay and disintegration of rocks, provided that no large scale transport of the loosened products is involved. The work of rainwash and wind, which is essentially erosional, is thus excluded" (A. Holmes, 1952).

It appears from the above definitions that weathering is essentially the breakdown of rocks due to chemical and mechanical processes at their places. The definition of weathering by B.W. Sparks highlights the above facts. According to him, "weathering may be defined as the mechanical fracturing or chemical decomposition of rocks by natural agents at the surface of the earth."

It is obvious that weathering involves two types of changes in the rocks e.g. (i) physical or mechanical changes, wherein rocks are disintegrated through temperature changes (heat factor), frost-action (frost factor), biological activities (biotic factor), and wind actions; (ii) chemical changes wherein rocks are decomposed through static water, oxygen, carbon dioxide and biological activities. Secondly, the breakdown of rocks occurs at the place of rocks (in situ). Thirdly, there is no large-scale transport of weathered materials except mass movement or mass translocation of weathered materials (rock-wastes) down the slope under the force of gravity. Weathering, thus, may be defined so as to include all aspects of the mechanism of breakdown of rocks as follows.

"Weathering refers to the breakdown or disintegration and decomposition of rocks in situ through mechanical and chemical changes in the rocks and their minerals effected by water, temperature, wind, different atmospheric gases and organisms provided that there is no large-scale transport of weathered products by denudational processes except mass movement of rockwastes (weathered products) down the slope under the impact of gravity. " - Savindra Singh.

Controlling Factors

The nature and magnitude of weathering differs from place to place and region to region. Weathering of rocks is affected and controlled by the agents of weathering, lithological and structural characteristics of rocks, height and slope factors. Besides, climatic conditions, topography and reliefs, flora and microfauna also affect different processes of weathering to greater extent. For example, disintegration of rocks is more effective in hot and dry region and in the regions where frost action is more dominant while chemical decomposition is more prevalent in hot and humid and temperate humid regions.

Composition and Structure of Rocks : Since weathering involves disintegration and decomposition of rocks and hence mineral composition, joint patterns, layering system, faulting, folding etc., largely affect the nature and intensity of weathering. For example, carbonate rocks (e.g. calcium carbonate, magnesium carbonate etc.) having more soluble minerals are easily affected by chemical weathering. Well jointed rocks are more subjected to mechanical disintegration. Rocks having vertical strata are easily loosened and broken down due to temperature changes, frost action, water and wind actions. On the other hand, the rocks having horizontal beds are more compact and are less affected by the mechanisms of disintegration and decomposition.

Nature of Ground Slope : Ground slope controls mechanical disintegration of rocks and mass movement of weathered products down the slope. The rocks in the regions of steep hillslope are easily disintegrated due to mechanical weathering and the weathering materials are instantaneously moved down the hillslope in the form of rockfall, debris fall and slide, talus creep etc. Instantaneous removal of weathering products allows continuous exposure of rocks to atmospheric conditions for further weathering. The regions of gentle and moderate ground slope are less affected by mechanical disintegration.

Climatic Variations : Climate is considered to be very important factor of all types of weathering. Climatic geomorphologists are of the view that each climatic type produces definite conditions for a particular type of weathering. For example, chemical weathering is more dominant in humid tropical areas because of more available water and high temperature. Because of abundance of moisture and high temperature leaching process and solution of rocks are more effective in the humid tropics. Mechanical weathering is less effective. On the other hand, mechanical disintegration of rocks is more dominant in the tropical and semi-arid regions. Rocks are weakened due to alternate expansion on heating during daytime and contraction on relative cooling during nights because of diurnal change of temperature. It may be pointed out that limestones are very weak rocks in humid climatic regions but they are relatively more resistant to weathering and erosion in hot desert climate. The rocks in dry temperate climates are more susceptible to mechanical weathering than chemical weathering because alternate expansion and contraction of cracks, fractures and joints of rocks due to alternate freeze and thaw of water accumulated in these cracks and fractures weaken the rocks. Rocks are least affected by mechanical disintegration in cold climate but chemical decomposition of rocks may be effective provided that the ground surface is not covered by ice cover for longer duration in a year. Both, mechanical and chemical weathering cease when the ground surface is covered by permanent ice sheets. Not only this, seasonal variations in climate of a region generate different conditions for weathering. For example, in monsoon climate rocks are subjected to mechanical disintegration during hot and dry summer months whereas chemical and biochemical weathering is more dominant during wet monsoon months.

Floral Effects : The nature of weathering is largely determined by the presence or absence of vegetations in a particular region. It may be pointed out that vegetation is partly a factor of weathering and partly a protector of rocks. In fact, vegetations bind the rocks through their network of roots and thus protect them from weathering and erosion but the same time the penetration of roots weakens the rocks by breaking them into several blocks. Dense vegetations protect the ground surface from the direct impact of sun rays. The micro-organisms associated with the roots of plants and trees encourage decomposition and disintegration of rocks through physico-biochemical weathering.

Processes of Weathering

Generally, weathering processes are conveniently divided into physical, chemical and biochemical processes but these are so intimately interrelated that it is practically difficult to isolate one process from the other. In fact, "no chemical weathering takes place without the production of physical stresses; disintegration of rock by thermal expansion probably does not occur in the absence of the chemical process associated with the presence of water; in the country of even sparsest vegetation chemical weathering is replaced in part by biochemical (process)" (R.J. Chorley, et al. 1985). Inspite of this limitation one has to divide weathering into physical weathering, chemical weathering and biochemical weathering on the basis of dominant agent of weathering and weathering process. The weathering agents are divided into 3 types as follows.

1. Physical or mechanical weathering agents
   - (i) Moisture and water
   - (ii) Frost
   - (iii) Insolation (temperature)
   - (iv) Wind
2. Chemical weathering agents
   - (i) Oxygen
   - (ii) Carbon dioxide
   - (iii) Hydrogen
3. Biological weathering agents
   - (i) Vegetation
   - (ii) Animals, mainly micro-organisms

Thus, weathering processes or simply weatherings are divided, on the basis of weathering agents, into 3 major types.

1. Physical or mechanical weathering
   - (i) Block disintegration due to temperature
   - (ii) Granular disintegration due to temperature
   - (iii) Block disintegration due to frost
   - (iv) Exfoliation or onion weathering due to temperature and wind
2. Chemical weathering
   - (i) Oxidation
   - (ii) Carbonation

(iii) Solution

(iv) Hydration

(v) Chelation

(vi) Hydrolysis

3. Biotic weathering and biochemical weathering

(i) Plant weathering

(ii) Animal weathering

(iii) Biochemical weathering

(iv) Anthropogenic weathering

Physical Weathering

The physical or mechanical weathering leads to fragmentation and breakdown of rock masses into big blocks and boulders, cobbles and pebbles, sands and silts and feldspar and mica minerals are chemically decomposed and clay is formed. Physical weathering may be defined as the disintegration of rocks due to temperature variations, frost action, wind action and unloading of confining super incumbent pressure. Though temperature variation is a key factor in physical weathering but pressure release, freeze and thaw of water and gravity also play major roles.

Block Disintegration due to Temperature Change : Temperature changes have been reported to have great impact upon many rocks but there are also some rocks which are least affected by temperature changes such as elastic sedimentary rocks (e.g. shales and sandstones) because the particles are separated by thin cementing laminae of silica. On the other hand, crystalline rocks, like granites, are more affected by temperature changes as particles are closely associated with each other and these particles expand and contract with increase and decrease of temperature respectively. It has been experimentally demonstrated that if the temperature of granite rocks is increased by 65.5°C, the rock contracts by 2.54 cm per 30.48 m distance. Contrary to this Black Welder in 1925 found no impact of temperature change on granite when he dropped granitic blocks into hot oil at the temperature of 200°C. It may be pointed out that contrasting results have been reported about the impact of temperature changes on the rocks.

The products of weathering in hot desert areas are different from those of more humid areas as they are coarser and deficient in clay and organic matter. Generally, it is accepted that the bare rock surfaces are heated during day time due to which their outer layers expand. During nights the rocks are cooled due to relative decrease in temperature which leads to contraction in the outer layer of the rocks. Thus, the repetition of expansion and contraction of outer rock layers due to diurnal range of temperature in the hot desert areas causes tension and stresses which introduce parallel joints in the rocks. The rocks, then, are disintegrated along these joints and broken big blocks of rocks are dislodged from the main rock mass and fall down the slope under the impact of gravity. This process of physical weathering is called block disintegration. It may be pointed out that block disintegration should not be considered as the result of only temperature changes, rather unloading of super incumbent load or release of confining pressure also helps in this process.

Granular Disintegration due to Temperature Changes : The coarse-grained rocks are more affected by shattering process in those hot deserts which are characterized by high range of daily temperature. If the rocks are coarse-grained and are of different colours, they absorb insolation differently. Thus, the different parts of the same rock mass receive and absorb different amount of insolation, consequently the different parts of the rocks are affected by differential expansion and contraction which cause stresses within the rocks due to which they are disintegrated into smaller particles. Such type of shattering of rocks is called granular disintegration which is more active in hot desert areas.

Shattering due to Rain Shower and Heat : The outer shells of the rocks are shattered due to sudden light showers in hot climatic regions mainly in hot desert areas. Griggs has remarked after experiments that small cracks are developed at the outer surface of the highly heated rocks when light drizzles suddenly strike them. It may be mentioned that Griggs' experiments involving purely thermal changes equivalent to diurnal change of 110°C over 244 years could not produce any change in the rock strength and thus could not cause any disintegration of rocks. In another experiment Griggs used water sprinkles to cool highly heated rocks instead of reducing the temperature. The result was imminent as the rocks developed cracks and surface spalling. This process works when there is sudden light showers in the hot desert areas. The highly heated rocks when struck by sudden drizzles develop numerous cracks. The repetition of this mechanism causes spalling and granular disintegration of rocks.

Block Disintegration due to Frost : Disintegration of rocks into large size blocks due to freeze and thaw of water is of common occurrence in the temperate and cold climatic regions. In fact, this process is more active in those areas which are very often characterized by alternate process of freezing and thawing of water mainly during night and day respectively. Frost action weakens the rocks in two ways e.g. (i) due to freeze and thaw of water between the particles of the rocks and (ii) due to freeze and thaw of water in the crevices and pore spaces. The more compact and highly consolidated rocks, like granites, are least affected by freeze-thaw actions while less compact and loosely consolidated rocks are more affected by frost actions, for example, sedimentary rocks being more porous are highly susceptible to the mechanism of weathering. Water present between the particles of porous rocks freezes during night due to fall of temperature below freezing point and thus expands due to increase in its volume by about 10 per cent and thaws during day time due to relative increase in the temperature and hence it contracts in volume by 10 per cent. This diurnal freeze and thaw cycle causes alternate expansion and contraction which introduce tension and stresses due to which rocks are disintegrated into smaller particles. This process, known as granular disintegration due to frost action, is an exceedingly slow process and rocks are least affected by this process.

Alternatively, alternate expansion and contraction of crevices, pores and cracks in the rocks due to diurnal freeze and thaw of water causes block disintegration of rocks wherein rocks are broken down into larger blocks which are dislodged from the main rock mass. When such process operates over the hillslopes of well jointed massive rocks, the dislodged rock blocks tumble down the slope in the form of rockslides and rock falls and collect at the base of the hillslopes.

The disintegration of rocks due to diurnal freeze-thaw cycles in the periglacial areas is called frost weathering or congelifraction which forms very interesting landforms like frostriven polygons. 5. Exfoliation due to temperature and wind- Exfoliation weathering, also known as onion weathering, refers to peeling off concentric shells of rocks due to combined actions of heat and wind in hot arid and semi-arid regions and monsoon lands. Exfoliation is more common over crystalline rocks. The outer shells of rocks become loose due to alternate expansion and contraction due to high temperature during day time and comparatively low temperature during night respectively and these loosened shells are removed (peeled off) by strong winds. Differential heating of outer and lower shells of a rock mass causes flaking. The solar radiation penetrates upto a few centimeters only in the rocks having low thermal l conductivity. Thus, the outer shells of such rocks expand more than the shells lying just below. This differential expansion of rock shells causes flaking wherein the thin rock sheets are detached from the rock mass. These detached rock sheets are later on removed by strong winds. Thus, sheets after sheets of rocks are peeled off and the rocks continue to be bare. Many of the granitic batholiths, which are exposed above the ground surface, are being continuously affected by exfoliation weathering. Kanke Dome near Ranchi city exhibits a fine example of such weathering process.

Disintegration and Exfoliation due to Unloading : The rocks, which are buried under thick covers of overlying rocks, are disintegrated when they are exposed to the surface due to removal of super incumbent load and consequent release of confining pressure. The removal of super incumbent load, very precisely known as unloading, may be effected through gradual denudation of overlying rocks. In fact, the burried rocks, when relieved of the confining pressure due to unloading, develop cracks and joints and ultimately breakup along these cracks and joints. Granites, massive sandstones, massive arkose, conglomerates and limestones are more affected by sheeting, spalling and block disintegration due to reduction in the confining pressure because of removal of super incumbent load by uplift and erosion.

Sheeting refers to the development of cracks and fractures parallel to the ground surface caused by removal of super incumbent load resulting into reduction of confining pressure. Such parallel cracks and fractures are developed in massive rocks such as granites and other igneous intrusives, quartzites and thickly bedded sandstones because of expansion of rocks consequent upon unloading of super incumbent load. R. H. Jahns (1943) has enumerated seven processes 'which cause sheeting in the rocks

(i) Tensional or contractional strains set up during cooling of an igneous mass,
(ii) Local or regional compressional stresses due to tectonic movements,
(iii) Insolation, with attendant daily and secular temperature changes,
(iv) Progressive hydration and formation of chemical alteration products in susceptible minerals,
(v) Mechanical action of fire, frost, and vegetation,
(vi) Diminution of primary confining pressure by removal of super incumbent load, and
(vii) Combinations of the above causes. Cambering process refers to fracturing of brittle sandstone beds along vertical joints due to expansion caused by unloading of super incumbent load and consequent release of confining pressure. G.W. Bain (1931) has reported the case of flying rock-sheets or spalls known as rockbursts in limestone quarries due to spontaneous mechanical rock expansion caused by unloading of super incumbent load. "As new faces are cut in the walls of (limestone) quarries the dense limestone expands, producing cracks parallel to the surface. In some instances quarries had to be closed down because of the danger of flying rock sheets or spalls" (C.D. Ollier, 1969).

The process of spalling refers to the development of platy rock fragments, lozenge shaped or irregular, in the rocks due to unloading of super incumbent load.

Other Types of Physical Weathering : Boulder cleaving refers to breaking and splitting of boulders of granites and basalts and complex boulders due to thermal expansion. Another type of insolation weathering is 'dirt cracking' wherein the boulders containing 'dirt' are fractured and split due to thermal expansion and contraction. Fire, mainly brushfire, also causes insolation weathering due to thermal expansion and contraction of rocks which cause exfoliation and thus numerous spalls and flakes of rocks are produced. Slaking weathering refers to the disintegration of rocks due to alternate wetting and drying of rocks wherein consequent expansion and contraction of rock shells result in the disaggregation of rocks. Disaggregation of rocks due to growth of salt crystals from solution is called salt weathering which generally occurs in hot arid areas. It may also be important in the rocks of coastal areas.

Chemical Weathering

Decomposition and disintegration of rocks due to chemical reactions is called chemical weathering wherein the minerals of the rocks weather away. Water vapour and water are the media which activate several types of chemical reactions within the rocks. Pure water, distilled water, is chemically inert but when it mixes with the atmospheric gases, mainly with CO_2, it becomes potent solvent. Oxidation, carbonation, solution, hydration, chelation, hydrolysis, base exchange etc. are the important chemical reactions which cause various chemical changes in the minerals of rocks which ultimately lead to decomposition and disintegration of rocks.

Solution : Solution is considered to be the first step in the chemical decomposition and disintegration of rocks. Solution refers to the dissolution of soluble particles and minerals from the rocks with the help of water in motion but a thin film of water around a solid particle also leads to chemical dissolution. Solution of rocks depends on the nature of rocks, solubility of rocks or solids and the ratio between the volumes of solvent (water) and the solids. Common salts are most soluble whereas carbonate rocks (limestones-calcium carbonates, dolomites-magnesium carbonates etc.) are of moderate solubility.

Limestones are more susceptible to solution process which depends on temperature, CO_2 (carbon-dioxide) content of water and pH of the solution. When rainwater mixes with atmospheric CO, it becomes active solvent and when it comes in contact with the carbonate rocks, such as limestones and dolomites, it dissolves the rocks through a set of chemical reactions occurring through various stages. The various stages of the chemistry of limestone solution may be presented in a simplified form as follows.

According to R.M. Garrels (1960) there are seven variables which control the equilibria involved in the solution of limestones

(i) Partial pressure of CO_2

(ii) $[H_2CO_3]$... carbonic acid

(iii) $[HCO_3^-]$...bicarbonate ion

(iv) $[CO_2^3]$...carbonate anion

(v) $[H^+]$ ... hydrogen ion

(vi) $[OH^-]$...hydroxyl ion

(vii) $[Ca^{2+}]$...calcium cation

Calcium hydroxide, $Ca(OH)_2$, is formed due to reaction of calcium oxides (C_2O) with water (H20) in the following manner of reversible exothermic reaction-

$CaO + H_2O = Ca\ (OH)_2$ ....eq.

Calcium carbonates, $CaCO_3$, is formed due to reactions of calcium hydroxide (Ca $(OH)_2$) with carbon dioxide (CO_2) in the following manner of reversible exothermic reactions

$Ca\ (OH)_2 + CO_2 \leftrightarrow CaCO_3 + H_2O$...eq

Carbonic acid (H_2CO_3) is formed when CO_2 is dissolved in water -

$CO_2 + H_2O \leftrightarrow H_2CO_3$ .... eq.

Carbonic acid is also dissociated into positive hydrogen ion and negative bicarbonate ion -

$H_2CO_3 \leftrightarrow H^+ + HCO_3^-$ .... eq.

Calcium carbonate (limestones) dissociates to limited extent in pure water into a metal cation (Ca^{2+}) and carbonate anion (CO_3^{2-}) during the process of dissolution in the following manner -

$$CaCO_3 \xrightarrow{\text{dissociation}} Ca^{2+} + CO_3^{2-} \text{......eq}$$

It may be pointed out that limestone can be dissolved in water only when it is transformed into calcium bicarbonate, Ca $(HCO_3)_2$, with help of carbonic acid (H_2CO_3) as follows

$CaCO_3 + H_2O + CO_2$ ® $Ca(HCO_3)_2$ ...eq

or

$CaCO_3 + H_2CO_3$ ® Ca $(HCO_3)_2$

The carbonate ion (CO_3^{2}) can react with water when it accepts proton from acid and then the carbonate becomes base (which accepts the proton). The ultimate reactions yield hydroxyl ions and thus the calcium carbonate becomes an alkaline substance as follows

$$CO_3^{2-} + H_2O \xrightarrow{\text{hydrolysis}} HCO_3^- + OH^- \text{...eq}$$

The actual quantity of limestone dissolved in water depends on temperature, CO_2 content of water, partial pressure of CO_2, pH of the solution and kinetics of reactions. The solubility of CO_2 is directly related to pressure (partial pressure) and is inversely related to temperature. In other words, the solubility of CO_2 increases and decreases with increase and decrease of partial pressure whereas it (solubility of CO_2) increases with decrease in the temperature and vice versa. On the other hand, the solubility of solids (say limestones) is directly related to temperature i.e. total solution of limestone in creases with increase in temperature and vice versa. More and more limestones can be dissolved in water either by increasing the temperature or CO_2 content of water or by decreasing the pH of the solution. The solution of limestones and dolomites gives birth to very interesting landscapes known as karst topography characterized by various solution holes (sink holes, swallow holes, uvulas, dolines and polje) and various types of caves and galleries.

Oxidation : The chemical process of oxidation simply means a reaction of atmospheric oxygen to form oxides. When water is mixed with oxygen its reaction with the minerals of the rocks forms hydroxide. In other words, the atmospheric oxygen after reacting with the rocks produces several types of oxides, iron oxide being the most important, which weakens the rocks to disintegrate. The oxidation of minerals of the rocks by gaseous oxygen becomes possible when oxygen is dissolved in water. Most of the iron bearing rocks commonly contain iron in ferrous state (Fe) e.g. major iron sulphide (pyrite, FeS_2), iron carbonate (siderite, $FeCO_3$), and various iron silicates.

When water mixed with atmospheric oxygen comes in contact with iron bearing rocks, the iron oxidizes to form ferrous oxides (FeO). Further oxidation of ferrous oxides produces ferric oxides (Fe_2O_3) or ferric hydroxides ($Fe(OH)_2$). The oxidation of iron-bearing rocks reproduces rusts in the following manner

$$Fe + H_2O + O_2 \rightarrow \underset{\text{(rust)}}{Fe_2O_3.3H_2O} \text{.......eq}$$

The rusting of rocks weakens them and ultimately the rocks are disintegrated. The ferric oxides and ferric hydroxides give red and yellow colours to many rocks and soils. The oxidation of iron-rich Vindhyan sandstones of the Kaimur Ranges and Rewa scarps (M.P.) has helped in the block disintegration of massively bedded and well jointed sandstone capping.

Carbonation : 'Carbonation is the reaction of carbonate or bicarbonate ions with minerals'. The process of carbonation is also known as 'solution' wherein atmospheric carbon dioxide after mixing with water forms carbonic acid ($H_2 CO_3$) which after reacting with carbonate rocks, say limestones ($CaCO_3$), forms calcium bicarbonate ($Ca(HCO_3)_2$) which is easily dissolved in water. The mechanism of solution of carbonate rocks has already been discussed above. The rainwater having dissolved carbon dioxide (CO_2 aq) percolates through the different horizons of the soils to reach underlying limestones. Thus, more and more organic carbon dioxide is dissolved in groundwater which then becomes a more active solvent because dissolution of more carbon dioxide produces more carbonic acids which dissolve more carbonate rocks after transforming calcium carbonates into calcium bicarbonates.

Hydration : The process of hydration is related to the addition of water to the minerals. The rocks after having absorbed water undergo the process of positive change of their volume. In other words, the volume of the hydrated rocks (rocks which have absorbed water) increases remarkably. Some times the increased volume becomes about twice the original volume. Thus, the increase in the volume of rocks due to increase in the volume of minerals causes stresses and strains in the minerals of the rocks which ultimately lead to physical disintegration of rocks. "Hydration is an exothermic reaction, and involves a considerable volume change which may be important in physical weathering exfoliation and granular disintegration. Hydration prepares mineral surfaces for further alteration by oxidation and carbonation, and enables the transfer of ions to take place with greater ease" (C.D. Ollier, 1969). The process of hydration changes feldspar minerals into kaolinite clays, the process being known as 'kaolinization'.

Hydrolysis : "Hydrolysis is a chemical reaction between mineral and water, that is between hydrogen (H) ions or hydroxyl (OH) ions, and the ions of the mineral" (C.D. Olier, 1969). In fact, the hydrolysis is that process wherein both the minerals of the rocks and water molecules decompose and react in such a way that new mineral compounds aıe formed. Silicate minerals are most affected by hydrolysis. This reaction starts immediately when a mineral comes in contact with water. The hydrolysis of magnesium silicate minerals (Mg_2SiO_4) in contact with 4 ionized water molecules ($4H_2O \rightarrow 4H^+ + 4OH^-$) takes place in the following manner—

$$Mg_2SiO_4 + 4H^+ + 4OH^- \xrightarrow{\text{hydrolysis}} 2Mg^{++} + 4OH^- + H_4SiO_4 \text{ (silica acid in solution)}$$

Hydrolysis of potassium feldspar (orthoclase, 2 $KAlSi_3O_2$) with carbonic acid (H_2CO_3) in water is perhaps the most common type of chemical weathering process wherein the end product of the reaction of potassium feldspar with carbonic acid in water is potassium and bicarbonate ions in solution. The mechanism of the hydrolysis of potassium feldspar is given below -

$2KAlSi_3O_2 + H_2CO_3 + 9H_2O \rightarrow Al_2S_2O_5(OH)_4+$

orthoclase + carbonic + water kaolinite, a clay

acid mineral

$4H_4SiO_4 + 2K^+ + 2HCO_{3^-}$

silicic acid + potassium and bicarbonate

in solution ions in solution

Chelation : According to D.S. Lehmann (1963) "chelation is a complex organic process by which metallic cations are incorporated into hydro carbon molecules". In fact, the word 'chelate' means a coordination compound in which a central metallic ion is attached to an organic molecule at two or more positions. In other words, chelation means 'holding of an ion, usually a metal, within a ring structure of organic origin' (C.D. Ollier, 1969).

We may safely say that chelation is a form of chemical weathering by plants. Plants extract minerals or say nutrients from the soils with the result mineral lattices are disrupted and crystal lattices are fragmented and thus mineral weathering takes place at a much faster rate.

The products of chemical weathering are classed under three categories.

(i) Solutes of sodium, potassium, calcium, magnesium etc., produced by the process of carbonation or solution of carbonate rocks, which are brought to the lakes and seas and are reprecipitated to form limestones', dolomites and other carbonate rocks.

(ii) Clays, derived from the weathering of feldspar and ferromagnesian minerals, form argillaceous sedimentary rocks like shales.

(iii) Mineral residuals, such as silica, unweathered feldspar and mica and other heavy minerals, form clastic sedimentary rocks such as sandstones.

Biotic Weathering

Plants and animals including man largely control the breakdown of rocks. It may be pointed out that in all types of weathering in all climatic regions biotic communities play some roles in one way or the other. This is why B.B. Polynov (1937) believed that completely sterile weathering was impossible. It may be mentioned that it does not mean that biotic communities always indulge in destructive work by disintegrating and decomposing the rocks but the burrowing animals definitely help in the transfer of soils from lower to upper and upper to lower horizons and thus the mixing of geo-materials activates weathering. Though vegetations protect the rocks by binding them through their roots but different types of acids (e.g. humic acids, bacterial acids, micro floral acids etc.) produced by them facilitate biochemical weathering. Recently, man has become the most powerful weathering agent because of the development of modern technologies. Biotic weathering, thus, is divided into 3 types e.g. (i) faunal weathering, (ii) floral weathering and (iii) anthropogenic weathering.

Faunal Weathering : The burrowing animals, worms and other organisms help in gradual breakdown of rocks or fragments thereof. Burrowing animals include gophers, prairie dogs, foxes, rabbits, jackals, termites, rats etc., which dug out burrows and tunnels in the rocks and unconsolidated geo-materials as their living places (homes). By doing so they weather the rocks and geo-materials to great extent. Small organisms play more important roles in rock and soil weathering. These organisms repeatedly mix up the soil materials and thus always expose fresh materials to weathering agents. They also help in moving the organic matter downward into the soil profiles and thus extend the weathering at greater depths which otherwise would have not been possible.

It is believed that there are about 1,50,000 creatures, big and small, in one acre of land and these organisms bring about 15 tonnes of soils at the surface from below every year. According to the estimate of Charles Darwin the soil organisms bring about 25.4 thousand kilograms of soil at the surface every year in the English gardens. Termites play very important role in sorting and rearranging the soil materials in the upper horizons of soil profiles in tropical regions. Termitaria are the evidences of soil weathering by termites. According to Ponomareva (1950) earthworms burrow to about 1.5 m and pass 10 tons per acre per year as a mean and 20 tons per acre per year as a maximum of soil materials. Rabbits, prairie dogs etc., destroy the soil structure and they obstruct the leaching and other horizon forming processes by constantly remixing the soil materials.

Floral Weathering : Weathering of rocks by vegetations takes place in two ways viz. (i) physical weathering and (ii) chemical weathering which is called as biochemical weathering, which will he discussed under separate heading. It may be pointed out that floral weathering does not take place independently rather it helps the physical and chemical processes of weathering. Larger plants affect and control weathering in a number of ways. (i) Cracks are widened by root penetration and consequent root pressure. (ii) Dense vegetation cover generates distinct microclimate at the ground surface. The soil atmosphere is largely affected by root respiration, humus content, increased moisture due to low rate of evaporation, increased content of organic CO_2, low temperature, all of which activate chemical weathering. It may also be mentioned that vegetations also protect the rocks and soils from weathering processes.

Anthropogenic Weathering : Man being a biological agent accelerates and decelerates the natural rates of weathering by many folds. The 'economic and technological man' lashed with modern technologies has become the most powerful weathering and erosion agent. Mining activities for extraction of minerals, blasting of hills and ridges by dynamites for road and dam construction and mineral extraction, quarrying for industrial (limestones for cement) and building material etc. result in such a fast rate of disintegration of geo-materials (rocks) that this may be accomplished by natural weathering processes in thousands to millions of years. Man accelerates the rate of weathering on hillslopes by modifying the ground surface through deforestation which reduces the mechanical reinforcement and cohesion of unconsolidated geo-materials and thus increases slope instability which causes slope failures and mass movement of materials down the slope in the form of landslides, slumping and debris fall and slides.

Biochemical Weathering

Biochemical weathering refers to decomposition and disintegration of rocks due to organic materials of both flora and fauna. A complex set of different biochemical processes such as cation root exchange, chelation, solution by root exudates and production of different kinds of organic acids such as humic acids, bacterial acids, microfaunal acids etc., produced by organic materials help in the decomposition and disintegration of rocks and soils.

Humic acids activate chelation and help in the decomposition of silicate minerals. Fulvic acids, humic acids derived from peat, play important role in decomposing rock minerals. Bacterial acids, including lactic, acetic, oxalic and gluconic, attack a wide range of rock minerals important being magnesium carbonate, calcium and magnesium silicates, feldspar and kaolinites. Bacterial acids also produce sulphides, oxidize iron and help in the solution of silica when the rocks are constantly submerged under water (perpetual waterlogging). Microfaunal acids such as oxalic and citric acids are produced by fungi and lichens. These acids weather silicate minerals and clays.

Chemotrophic bacteria manufacture sulphides and remove silica in the tropical soils and help in the carbonate mineralization in caves. The colonization by blue-green algae forms desert varnish and mobilizes ferrous irons and help in the concentration of oxides on rock surfaces. Microorganisms also form varnish incrustations on rocks. Lichens introduce the alteration of minerals composition of rocks both mechanically and chemically. The organic carbon dioxide produced by plants accelerates the rate of carbonation on carbonaterocks e.g., limestones and dolomites.

Geomorphic Significance

Production of Rockwastes : Rocks are disintegrated and decomposed and ultimately are broken down into smaller pieces due to the operation of different weathering, chemical weathering, biotic weathering and biochemical weathering. Thus, different weathering processes produce immense volume of rockwastes or weathered materials. These weathered materials lying over the unweathered fresh rocks are called regoliths. The depth of weathered rocks from the ground surface to the unweathered fresh rocks is called weathering zone. The depth of weathering zones varies from place to place and from region to region depending mainly on the depth of water table of groundwater and the duration of weathering. The weathered materials are very important economically because they help in the process of soil formation, they expose minerals etc. Weathering generates mass movement of rockwastes down the hillslope and thus causes damage to human settlements in the foothill zones, causes obstructions in the river flow and thus forms lakes (by damming the rivers through debris fall).

Weathering helps Erosional Processes : Weathering loosens the rocks by disintegrating and decomposing them and thus paves the way for erosional processes to operate easily. Different agents of erosion like running water (rivers) in humid regions, wind in hot arid and semi-arid regions, glaciers in cold regions and sea waves operating along coastal zones obtain these weathered materials and move them to other places. The rapid rate of weathering due to mass felling of trees (deforestation) has accelerated the rate of erosion of nude rocks of the hill ranges with the result most of the rivers have become overloaded and sluggish because millions of tonnes of eroded sediments are reaching the major rivers every year. For example, Garhwal and Kumaun Himalayas and other parts of the Himalayas have been extensively deforested and thus the weathered rocks have increased the rate of fluvial erosion, consequently most of the Himalayan rivers like the Yamuna, the Ganga, the Ghaghra, the Kosi etc., have become overloaded in the plains due to supply of huge volume of sediments every year. This process has caused rapid rate of siltation of river beds of major alluvial rivers of north India and the resultant siltation has increased the frequency and dimension of recurring floods.

Lowering of Surface : Continuous removal and transfer of weathered materials through different processes of masstranslocation of rockwastes such as landslides, debris slides, rockfall, rockslides, talus creep etc., and by the agents of erosion causes gradual lowering of the height of the affected area.

Evolution of Landforms and their Modifications : Differential weathering helps in the evolution of different types of landforms. Weathering plays important role in the development of stone lattice (in hot deserts), tors, buttes, talus cones, talus fans, sandstone anvils etc. It may be pointed out that weathering and erosion go hand in hand and thus it is not wise to separate the inseparable, so it is difficult to ascertain the quantum of work done by weathering and erosion in the development of a particular type of landform.

Mass Movement Rockwastes

The study of mass movement of rockwastes involves the analysis of meaning and concept, classification, causes, and geomorphic significance of mass movement or mass translocation of rock wastes.

Meaning and Concept : Disintegrated and fragmented rock materials due to mechanism of weathering processes (mechanical, chemical, biotic and biochemical) are called rockwastes. Generally, movement of rockwaste enblock down the hillslope is called mass movement of rockwaste or simply mass movement. 'Mass movement is the detachment and downslope transport of soil and rock material under the influence of gravity. The sliding or flowing of these materials is due to their position and to gravitational forces, but mass movement is accelerated by presence of water, ice and air. This definition of mass movement permits consideration of the movement of earth materials at all scales and at all rates' (R.J. Chorley, et. al, 1985). It is evident from the above definition that mass movement includes both, detachment of rock materials and their downslope transport enblock. 'The collective term for gravitational or downslope movements of weathered rock debris is mass-wasting. The term implies that gravity is the sole important force and that no transporting medium such as wind, flowing water, ice or molten lava is involved. Although flowing water is excluded from the process by definition, water nevertheless plays an important role in mass-wasting by over steepening slopes through surface erosion at their bases and by generating seepage forces through groundwater flow' (A.L. Bloom, 1978).

If we look at the aforesaid two definitions of R.J. Chorley et. al and A.L. Bloom it appears that the term mass movement is more sound and appropriate than mass wasting to describe enblock downslope transport of weathered materials ranging from very fine (soils) to very coarse and large sized rock materials (boulders). In fact, the definition by R.J. Chorley and others is comprehensive one because it includes both the aspects of mass movements, viz. detachment of rock materials and their downslope transport whereas Bloom's mass wasting describes only the process of downslope transport of weathered rock debris.

Emphasising the significance of tectonics in mass wasting (may be rock disintegration) and mass movements R.J. Chorley and others (1985) have remarked that, 'the relation between mass wasting and tectonics is a relatively clear one. Where rocks are shattered and relief is high, this is where mass movement is common and, in fact, the denudation of high mountains may ...... be the result of mass wasting rather than fluvial or glacial process'.

It is, thus, evident that mass movement of rock wastes includes the mechanisms of detachment of rock materials through different weathering processes, and enblock downslope transport of weathered rock debris by gravity force without any medium of transport (e g running water, wind, sea waves, glacier etc.) except some lubricating role of water or ice. The rock debris coming through mass movement are deposited at the foot-hill zone as scree or talus. The deposit of large boulders in conical shape is called talus cone. It is, thus, apparent that the most significant stimulating factor of mass movements is gravity force.

The Classification : A wide range of variations in terms of rate, direction and type of movements is noted in mass movements in different places having varying environmental conditions. It is generally believed that mass movement of rock wastes occurs suddenly and instantaneously and hence all mass movements cannot be witnessed by man. But in reality mass movements have long preparatory period and there are certain precursor events which herald the occurrence of mass movements but these are generally unnoticed. It may be mentioned that most of mass movements occur in mountainous areas and hence it is not possible to notice the precursor events such as restlessness of animals, deserting of hives by bees etc. 'Hence, if a landslide comes as a surprise to eyewitnesses, it would be more accurate to say that the observers failed to detect the phenomena which preceded the slide' (R.J. Chorley et. al, 1985). Mass movements are generally classified on the basis of causative factors e.g., rate of movement, direction of movement, type of movement, lubricating substance e.g. water, ice etc.

The direction of mass movement of rockwaste down the slope may be (i) vertical, (ii) lateral, and (iii) diagonal. Based on direction mass movement may be divided into vertical movement, lateral movement and diagonal movement of rockwaste. Vertical mass movement is further divided into (a) rockfall, (b) collapse earthfall. Lateral mass movement includes (a) block slide, (b) spread, (c) cambering, (d) sackung etc. Diagonal mass movement is divided into (a) soil creep, (b) rockcreep, (c) talus creep, (d) rockslide, (e) debris slide, (f) slump, (g) debris flow, (h) mud flow, (i) solifluction, (j) avalanche etc. R.J. Chorley et. al (1985) have presented exhaustive classification of mass movement-mass wasting phenomena on the basis of direction of movement, type of movement and presence of transporting agent as given below

Table : Mass Movement-Mass Wasting Phenomena

Direction of movement	*Vertical*	*Lateral*		*Diagonal*			
Type of movement	Fall	Subsidence	Slide	Spread	Creep	Slide Flow	
Presence of transporting agent	No	No basal layer or on sliding surface	Minor in in basal layer	Moderate	Minor moderate	Minor to	Major
Types of mass movement mudflow	Rockfall, earthfall, topple	collapse	Block slide settlement	Spread cambering	soil creep, rock creep, talus creep	rock slide debris slide, soil slip, slump	soli fluction rock glacier, rock avalanche

Source : R.J. Chorley et. al, 1985.

Based on the rate of movement and water content mass movements are classified in 3 types:

(1) *Large-scale Rapid Slide of Rock Waste* : Water is needed as lubricating agent for such type of mass movement. Landslide is the typical example of this type

(2) *Slow Flowage of Rock Waste and Weathered Debris* : Partial saturation of rock debris is required for such mass movement and hence moderate quantity of water is needed as lubricating and stimulating agent. Rock creep, soil creep, solifluction etc. are typical examples of this type.

(3) *Rapid Flowage of Weathered Debris* : Sufficient quantity of water is needed as lubricant. Earth flow, mudflow etc. are representative of this type of mass movement.

A generalized classification of mass movement of rock wastes is presented as follows.

Table : Classification of mass movement

1. Very rapid movement
   (no water is required)
   (i) Landslides
      (a) slump
      (b) debris slides
      (c) debris fall
      (d) rock fall
      (e) rock slides
2. Slow movement (flowage or slide) (little water is required)
   (i) Rock creep
   (ii) Soil creep
   (iii) Solifluction
3. Rapid movement (flowage or slide) (enough water is required)
   (i) Earthflow
   (ii) Mudflow
   (iii) Sheetwash

On the basis of direction and type of movement the following types of mass movement of rock waste may be identified (simplified scheme of R.J. Chorley et. al, 1985)—

Table : Classification of Mass Movement

Vertical Movement :—

(A) Fall (of earth materials from very steep slopes like steep scarps and cliffs) On the basis of materials
   (a) rockfalls
   (b) earthfalls (of alluvia, soils, colluvia)
   (c) debrisfall (soils, alluvia, colluvia, vegetation and human structures)
   (d) topple (rotational fall of rock slabs, or of earthen material)

(B) Subsidence (of the ground surface)
   Sinking

(a) collapse (of roofs of underground caves or cavities or lava tubes)

(b) settlement (collapse of ground surface due to withdrawal of water, crude oil etc.)

Lateral Movement :—

(A) Slides (sliding) (movement of materials along a horizontal fracture or interface between two rock strata of varying resistance e.g. sandstone- shales or limestones-shales)

(a) block slide (block glide) (downslope movement of a single large block of massive rock on such a surface which has been lubricated by water)

(B) Spreading (lateral displacement of a series of rock blocks (multiple blocks) or mud block downslope)

(a) cambering (draping of sedimentary units)

(b) sackung (lateral spreading away from anticlinal crests)

Diagonal Movement :—

(A) Creeping (downslope movement of earthen materials at slow velocity)

(a) soil creep (movement of moistened soils downslope)

(b) rock creep (movement of rock upon rock)

(c) talus creep (rearrangement of scree and downslope movement)

(B) Slide (rapid rate of downslope movement of large quantities of debris of varying sizes) (on the basis of materials)

(a) rock slides

(b) debris slides

(c) soil creep

(d) slumping (movement of tine materials along a curved plane)

(C) Flows (dominant role of water, downslope transport of water-soaked fine debris)

(a) earthflow

(b) debrisflow

(c) mudflow

Table : Simplified classification of mass movement (Land Slides)

1. Fall

(a) Rock fall

(b) Debris fall

(c) Earth fall

2. Slides
   (a) Slump
      (i) rock slump
      (ii) debris slump
      (iii) earth slump
   (b) Slides
      (i) rock slides
      (ii) debris slides
      (iii) earth slides
3. Topples
      (i) rock topples
      (ii) debris topples
      (iii) earth topples
4. Flows
   (a) Rock flow
   (b) Soil flow
      (i) debris flow
      (ii) earth flow
5. Lateral Spreads
      (i) rock spreads
      (ii) debris spreads
      (iii) earth spread

Mass Movement Factors : Any sort of mass movement of weathered debris with any rate whether on hillslope or valley side slope depends on the ratio between shearing forces (simply known as stress) and resistance of materials to shearing forces (i.e. shearing resistance of materials) as follows—

$$Fs = \frac{\text{strength or shearing resistance of materials}}{\text{magnitude of shearing forces}}$$

where Fs = factor of safety

When the quotient of shearing resistance of materials (simply strength of materials) and magnitude of shearing forces i.e. safety factor (Fs) is less than 1.0 (i.e. when magnitude of shearing forces of hillslope or valley side slope exceeds the shearing resistance of materials resting on slopes) materials begin to move downslope and thus mass movement of weathered debris occurs. It is apparent that mass movement may occur when either shearing forces increase or shearing resistance of materials decreases. It may be pointed out that either of the two processes (increase in stress and decrease in resistance of materials to stress) may operate or both the processes may operate together.

Based on this corollary D.J. Varnes (1978) classified the factors which control mass movement of rockwastes in two broad categories and many subcategories—(1) factors which increase shearing forces (shear stress) and (2) factors which reduce resistance of materials to shear stress.

Table : Factors of Mass Movement

Category I

Factors which increase the shear strength

(External Processes)

1. Removal of lateral support (under cutting steepening of slope)
   (A) (Natural)
      (a) stream erosion
      (b) glacial erosion
      (c) marine erosion by sea waves
      (d) weathering
      (these factors lead to removal of lateral support)
      (e) previous rockfall or slide, subsidence or faulting (these factors steepen the slope)
   (B) (Anthropogenic factors)
      (a) Construction of quarries, pits, canals, roads
      (b) alteration of water levels in lakes and reservoirs
2. Surcharge (loading of slope)
   (a) (natural) weight of rain, snow (anthropogenic) water from pipelines, sewers, canals
   (b) accumulation of talus
   (c) vegetation, trees
   (d) seepage pressure of percolating water
   (e) (anthropogenic) construction of fill, waste piles, buildings

3. Transitory earth stress (endogenetic processes)
   (a) earthquakes
   (b) vibrations, blasting, traffic
   (c) swaying of trees in wind
4. Removal of underlying support
   (a) undercutting by rivers and waves
   (b) solution at depth, mining (anthropogenic)
   (c) loss of strength of underlying sediments
   (d) squeezing out of underlying plastic sediments
5. Lateral pressure
   (a) water in cracks
   (b) freezing of water in cracks
   (c) swelling (hydration of clay)
   (d) mobilization of residual stress (pressure release)

Category II

Factors which decrease (reduce) the shear strength of materials

1. Weathering and other physicochemical reactions
   (a) softening of fissured clays
   (b) physical disintegration of granular rocks (frost action, thermal expansion etc.)
   (c) hydration of clay mineral causing decrease in particles cohesion, swelling
   (d) base exchange (changes in physical properties)
   (e) drying (desiccation) of clays and shales (racking, loss of cohesion)
   (f) removal of cement by solution
2. Changes in intergranular forces due to water content (porewater pressure)
   (a) saturation
   (b) softening of material
3. Changes of structure
   (a) fissuring of shales and consolidated clays
   (b) remoulding of loess, sand and sensitive clay

4. Organic
   (a) burrowing animals
   (b) decay of roots

Source : D.J. Varnes, 1978, in R.J. Chorley et. al, 1985

Recently, man has emerged as a significant factor of mass wasting and mass movement in almost all of the environmental conditions. His activities (e.g. deforestation for commercial wood and increase in agricultural land; construction of roads, dams, reservoirs; urbanization on fragile hillslope; manipulation of rivers, coastal areas etc.) destabilize hillslopes as well as valley side slopes and accelerate the process of mass wasting and mass movement and increase frequency and magnitude of different mechanisms of mass movement. Increased deforestation, cultivation on cleared hill slope, construction of roads and reservoirs in the Himalayas have made the mountain ecosystem more fragile and vulnerable to increase frequency and magnitude of different types of mass movement.

Types of Mass Movements

Landslides : It may be mentioned that generally all types of mass movements of rock wastes including soils and ice are collectively called as landslides which are variously classified on different bases i.e., direction of movement, type and rate of movement, nature of materials, presence or absence of lubricants etc. On an average, landslides (downslope movement of different types of debris enblock) are divided into five major categories e.g., fall, slide, topple, flows and lateral spreads. On the basis of nature of materials these are further subdivide into several types.

Falls : Instantaneous fall of weathered rock materials including large blocks from steep hillslopes or earthen materials from steep and cliffed valleysides of streams under the influence of gravity, is called fall. The size of rock fragments depends on the size and pattern of rock joints. This type of movement involves vertical displacement of materials without water. The velocity of fall is greatest of all other types of mass movement. According to A.L. Bloom, 'fall is a distinct landslide process, but it is rarely independent of subsequent events.' On the basis of materials fall is subdivided into rock fall, debris fall and earthfall.

'Rock falls are relatively small landslides confined to the removal of individual and superficial blocks from a cliff base' (M.J. Selby, 1982). Rock fall is facilitated by granular and block disintegration of rocks under the processes of mechanical weathering and limited action of oxidation in sandstones. According to M.J. Selby (1982) 'most rockfalls are promoted by hydrofacturing, stress release, the wedging action of tree roots, and other weathering processes. ... a common cause of rock falls is undercutting of a face by streams or the more rapid weathering of an underlying weak rock such as shale or mudstone.'

The frequency of rock falls depends on certain environmental conditions such as aridity/ humidity factor, lithological and structural characteristics of rocks, nature of slope and vegetation etc. In humid areas rock falls are very common features but in hot arid areas they are of very rare occurrence. Debris fall involves rapid rate of fall of weathered rockmaterials (which are finer than the materials involved in rock fall) downslope (it may be hillslope or steep valley side slope of streams) from great height. The fallen materials collect at the foot-hill or cliff base and form small mounds and ridges. Earthfall involves downslope movement of finer materials than debris fall.

Slides : Slides, very often known as landslides among general public, are most significant of all types of mass movements. 'Mass-wasting wherein a mass of rock or weathered debris moves downhill along discrete shear surfaces is defined as a slide' (A.L. Bloom). It may be pointed out that slides involve downslope displacement of both types of materials-weathered rock materials and soils. 'Slides in rock or soil are characterized by movement above a sharply defined shear plane. In rocks such as slate, schist, and many sedimentary formations the shear plane follows a structural plane within the rock such as a plane of foliation or bedding-and it is often straight' (M.J. Selby, 1982). Slides are promoted by a host of controlling variables such as nature of slopes (vertical and cliff slope is essential for slides), moderate lubrication by water, earth tremors, gravity, vertical and steeply inclined rock beds, base removal etc. Slides are more frequent in certain locations having favourable condition viz. (1) steep hillslope or steep valleysides of streams, (2) fault scarps, (2) rejuvenated fluvially eroded valleys, (4) sea coasts, (5) alluvial river valleys, (6) degraded hills and mountains (due to deforestation, road construction, settlement expansion etc.).

On the basis of nature of materials, direction and rate of movement (intensity) slides are divided into (1) slump (which is further divided into rock slump, debris slump and earth slump), (2) rock slides, (3) debris slide, and (4) earth slide.

(1) Slumping involves intermittent sliding of rock fragments, rock blocks or soils downslope along a curved plane caused by rotational movement and displaced blocks (whether rock blocks or soil blocks) cover very short distance. Slump is promoted by undercutting of slope base (with hillslope or valley side slope of streams) by streams, seawaves (in case of coast land) and by human activities (quarrying). In fact, 'slump is the form of slide most common in thick, homogeneous, cohesive materials such as clay. The surface of failure beneath a slump block is spoon-shaped, concave upward or outward' (A.L. Bloom, 1978).

Slumping of alluvial deposits of valley sides of alluvial rivers of north India through undercutting of valley sides by hydraulic action of the streams during wet monsoon period is of common occurrence. Slumping is consuming a large chunk of rich agricultural lands every year along the Ganga valley in U.P. and Bihar. Based on the nature of materials involved slump is subdivided into rock slump, debris slump and earth slump.

(2) Rock slide (also known as rock glide or block glide) is most significant of all types of slides wherein large rock blocks slide down the hillslope. 'Rock slides may be very large and catastrophic in mountain regions where the large available relief permits accelerations of rock debris to velocities as great as those of rock falls and rock avalanches' (M.J. Selby, 1982). Rock slides involve rapid movement of materials downslope. Some times, the velocity is so high and mass of materials is so enormous that 'rock slides can be dramatic forms of sliding mass-wasting if large masses of unweathered rock slide downhill along a sloping joint or a bedding surface' (A.L. Bloom, 1978). The Cross Ventre Slide of 1925 in Wyoming, USA and Turtle Mountain Slide of 1903 in Alberta, Canada, are typical examples of devastating landslides. The very massive landslide (rock slide), which occurred in the north-western side of Naini Lake (Nainital, U.P.) in 1884, was so enormous that the debris filled a sizeable portion of the lake.

(3) Debris slide is more extensive and occurs at larger scale than slump but there is little amount of water. Debris slide is promoted because of two basic factors-(1) saturation of rocks due to water, and (2) sudden downslope movement of unconsolidated mantle rock. The materials involved in debris slide is a mixture of soils and rock fragments (boulders). The debris collects at the foot-hill or the base of the valleys and forms interesting morphological features.

Flow : Diagonal downslope movement of rock fragments and soils along sliding plane with enough water is called flow (which is further divided into solifluction, debris flow, mud flow, earthflow, rock avalanche etc.). Flow involves downslope rapid movement of rock debris or soils saturated with water like viscous fluid. 'Dry flows in sand or silt are known, but most flows are saturated with water. Rates of movement are greater than for creep but range from imperceptibly slow to tragically rapid mud flows and avalanches. Flows typically move as lobes or tounges' (A.L. Bloom, 1978).

Debris flow involves downslope movement of enormous amount of viscous soils and boulders either separately or mixed together, and occurs mostly along river valley sides. The difference between debris flow, earth flow and mud flow is related to size of particles and amount of water. The size of particle decreases from debris flow to mudflow. 'The three terms form a series of progressively higher water content (i.e. water content increases from debris flow through earth flow to mud flow) but are often used interchangeably. Debris flows have 20-80 per cent particles coarser than sand sizes, whereas earth flows and mud flows are 80 per cent or more mud and sand. Mud flow is the most liquid "end member" of the series' (A.L. Bloom, 1978). Debris flow occurs mostly due to availability of water, presence of loosely deposited soils and fine rock materials, lack of vegetation cover, clay minerals in the soils, unstable slope, undercutting of slope (valley sides) by streams, earth tremors etc. 'Debris flows range in size from a few meters to over 1000 meters in width and may be tens of meters thick in places; more commonly they are 1 to 5 m thick' (M.J. Selby, 1982). Debris flow is most common on gully heads in the riverine tracts of major alluvial rivers.

Earth flow is promoted by excessive water received mostly through rainfall so that the materials are oversaturated. Earth flow is more common on planar hillsides or valleysides having alluvium, rich in clay minerals.

Debris flow of volcanic materials saturated with water on volcanic cones is called lahar. Heavy downpour mixing with falling volcanic dusts causes enormous mud flow as lahar on the steep slopes of volcanic cones which inflicts great damage to human health and wealth. For example great lahar created on the steep slopes of Kelut Volcano in Japan in 1919 killed 5500 persons.

Mud flow differs from earth flow in that former may be noticed by the observer while the latter cannot be noticed because earth flow is not very common. The water content is more in mud flow than in debris flow and earth flow. Mud flow is most common along valleysides of alluvial rivers and the debris (mud) so produced is transported by the rivers. The necessary conditions which promote mud flow include (1) steep and vertical slope, (2) presence of unconsolited materials on the upper surface so that these, when mixed with water, become viscous fluid and slippery, (3) intermittent supply of sufficient water as lubricant, and (4) absence of vegetation. Based on these factors Elliot Blackwelder (1928) considered arid regions as most favourable for mud flow. C.F.S. Sharpe (1938) has divided mud flow into three categories on the basis of spatial characteristics e.g. (1) mud flow of arid regions, (2) Alpine mud flow, and (3) volcanic mud flow.

Creep : Very slow and imperceptible downslope movement of materials (colluvium) is called creep. On the basis of materials involved in such movement creep is divided into (1) soil creep (fine weathered rock debris as well as soil) and (2) rock creep (unweathered joint blocks). It may be pointed out that the rate of movement of materials (colluvia) under creep is so slow (a few millimeters per year) that it becomes practically difficult for the observers to notice it.

Soil creep is also called as solifluction which occurs in a variety of climatic conditions ranging from tropical humid to periglacial climates. The process of debris movement in periglacial regions has been variously defined and a number of terms have been suggested. First J.G. Anderson (1906) proposed the term solifluction (solum-soil, fluere flow) for slow movement of debris, soaked with water, form higher to lower slaps. Solifluction term was replaced by congelifluction of J. Daylik (1951) to incorporate only soil-flow in the periglacial climate having permafrost below an active layer. K. Bryan (1946) used the term cryoturbation which included all types of mass movement of regoliths under periglacial environment. Recently, gelifluction is used in place of congelifluction.

Rock creep involves downhill movement of rock debris having relatively great depth (upto 300 m) but the movement is very slow and ranges between one meter to ten meters per year. 'It is distinguished from soil creep by its great depth and isolation from daily and seasonal climatic conditions, and from land sliding by the lack of a single clearly defined failure plane and slow rate of deformation' (M.J. Selby, 1982). The following conditions promote rock creep-deformation of rocks through bending, folding, bulging, fracturing, spreading; distortion and buckling of inclined rock beds of varying resistance, mechanical disintegration of rocks etc.

Topographic Expressions

Different types of mass wasting and mass movement create distinctive morphological features on hillslope and river valley sides and coastal lands. It may be pointed out that on one hand there is wide range of variation in mass movement because of varying controlling factors and conditions, there is almost uniformity in the resultant morphological features. The topographic features produced by mass movement say landslides include scars, ripple marks, terraces, meander widening etc.

7

Developmental Stages

Denudation chronology simply means reconstruction of denudational history of a given region. This is primarily based on 'historical or chronological approach of landform study which involves the basic concepts 'that there is sequential change in landforms through time', 'principles of uniformitarianism,' 'cyclic nature of earth's history', 'palimpsest topography' and Davisian model of 'cyclic evolution of landforms'. The primary goal of this approach is to reconstruct the chronological history of denudation of a given region known as denudation chronology and 'to identify, date and interpret planation surfaces developed in past cycles and subcycles of erosion' (R.J. Small, 1970).

Denudation of Chronology

Historical approach or denudation chronology approach involves description of landform evolution through successive stages of geological time or say cyclic time involving longer geological time and larger spatial scales. 'In this type of analysis the emphasis is placed on the historical development of the landscape, based on the cyclic concept of Davis, on the assumption that evidence of the past character of the landscape is still apparent in its present form' (C.A.M. King, 1966). In fact, denudation chronology is based on the concept of 'palimpsest topography' which means such a surface which bears the imprints of geomorphological processes during past geological periods after partially erased initial imprints (features) in the beginning. Palimpsest refers to that manuscript which has been written, erased, and rewritten several times. Similarly, palimpsest topography represents complex topographic features of a region which have been written (characterized by topographic features) by geomorphological processes, erased (previous geomorphological features partially destroyed by succeeding processes) and rewritten (production of new reliefs on older surfaces) several times.

An attempt is made to reconstruct the geomorphic history of the region concerned on the basis of present and remnant (relict) landforms following the dictum of 'present is key to the past'. This method of landform study is popularly known as denudation chronology (denudational history of a given Legion). 'The principal objective (of this method) is to identify, date and interpret planation surfaces developed in past cycles and subcycles of erosion' (R.J. Small, 1970) on the basis of evidences of drainage development, river capture, relic surfaces and past tectonic events. The degree of precision of landform analysis rests on deductive power of the researcher and level of qualitative and quantitative description of relic features.

The denudation chronology approach suffers from certain perceptible weaknesses. This approach is highly deductive because unknown events and their responses are described on the basis of very limited known information and evidences. In fact, the past geomorphic history is reconstructed on the basis of very small parts of the earlier landforms. 'An important criticism which has been levelled against the denudation chronology approach is that it succeeds in explaining directly only very small parts of the existing landscape, namely the fragments of former (erosion) surfaces which have been dissected and almost totally destroyed in some cases by more recent erosion' (R.J. Small, 1970). Secondly, historical approach is highly deductive and speculative because the old erosion surfaces and remnant forms have been so greatly modified by subsequent processes that it becomes difficult or say impossible to find out their original forms and initial heights. Thirdly, the dating of erosion surfaces is also highly speculative as valid geological evidences are not available.

Planation Surfaces

Meaning : The almost plain topographic surfaces having undulating ground surface and remnant low reliefs caused by dynamic wheels of denudational processes and cutting across geological formations and structures are generally called erosion or planation surfaces. Erosion surfaces form significant elements of landscape of a given region and provide prominent clues for the reconstruction of denudation chronology (erosional and depositional history) of that region. R.J. Small has commented that 'it is perhaps unfortunate that by constant usage geomorphologists have given the term 'erosion surface' a very limited and specialised meaning. It is, in fact, almost universally used to describe only flat or near-flat erosional plains, formed very close to a and resulting from cycles of erosion that have reached well beyond the stage of youth and in many instances into the stage of old age. Thus, peneplains, peneplains, pediplains and planes of marine erosion are all erosion surfaces in the accepted sense of the term' (R.J. Small, 1970). Besides, etch plains, cryoplains (by periglacial processes) etc. are also major erosion surfaces. Besides, there are some minor erosion surfaces viz. valley side benches (terraces), river terraces, marine benches (platforms), marine flats, marine terraces, raised beaches etc.

It may be mentioned that erosion or planation surfaces widely differ from rock-cut structural benches, as erosion surfaces cut across geological formation and structures while structural surfaces are structurally controlled. For example, if horizontal soft (weak) rock beds overlie horizontal resistant (hard) rock beds, soft rock beds are more or less uniformly eroded and thus are almost entirely removed and underlying relatively hard rock beds are exposed to environmental processes. The surface of such lithological formation is called structural surface. It is, thus, apparent that structural surfaces are formed due to removal of overlying weak strata by denudational processes and parallel to the exposed underlying strata and their dip angle. On the other hand, erosion surfaces are formed due to erosion of different rock types (soft and hard) and different geological structures (folded, faulted, uniclinal etc.) alike 'Strictly speaking, of course, all parts of the land surface that are not directly depositional in origin are surfaces of erosion. Thus the dip and scarp-slopes of a cuesta, although owing their forin in large measure to the influence of geological structure, are 'surfaces' produced by weathering and erosion' (R.J. Small, 1970).

Identifying the Surfaces

In spite of the fact that identification and recognition of erosion surfaces in the field is a difficult and tedious task, advocates of Davisian cyclic model of landscape development and denudation chronology have made commendable attempts for solution of this problem. It may be pointed out that it is very difficult for erosion surfaces to be found very close to present sea-level (base level of erosion) because, after their formation (peneplanation), these have been largely affected by tectonic movements (uplifting, subsidence, upwarping etc.). Most of the erosion surfaces have been uplifted and thus there is every likelihood that they must exist relatively at higher elevations. If the erosion surfaces are found close to the present sea-level, it clearly means that it has experienced subsidence resulting in the lowering of its height. It is also important to note that sea-level in itself is not stable rather it has experienced several phases of positive (rise) and negative (fall) changes due to climatic changes (glacial and interglacial ice ages) and tectonic movement (upliftment and subsidence of coastal land and sea floor) in the past geological and geomorphic history of the earth e.g. Carboniferous and Pleistocene ice ages punctuated by interglacial periods. Even Quaternary period registered several phases of sea-level changes which effected changes in base levels of erosion. 'At the onset of Quaternary era, the sea-level was some 600 ft (182.88m) higher than present; and some authorities have even suggested that at some stage in the Tertiary it may have stood at a height of nearly 2000 ft O.D. (410 m). There is indeed a case for regarding most landforms below 600 ft (182.88 m) as of wholly Quaternary origin, and those above 600 ft as of Tertiary age though some glacial and periglacial modifications of these must of course be envisaged' (R.J. Small, 1970). It may be mentioned that post-Quaternary true erosion surfaces are difficult to be found because these could not be formed due to lack of sufficient time required for the completion of such surfaces. That is why erosion surfaces younger than Tertiary era are not found but partial erosion surfaces might have developed.

It is, thus, evident that no post—Tertiary erosion surface may be possible. The surfaces may be of Tertiary period or older than Tertiary. If this is the case, the erosion surfaces might have undergone substantial changes due to longer period of their formation and existence. The changes might have been effected by erosional and depositional processes, tectonic movements (upliftment and subsidence), sea-level changes causing transgressional and regressional phases and to some extent by anthropogenic factors in localized areas.

The identification and determination of erosion surfaces are accomplished on the basis of certain morphometric techniques e.g. altimetric frequency histograms and curves and superimposed profiles and field checks.

The frequency maxima of spot heights, summit levels, benches and shoulders as revealed by frequency histograms and curves of these variables separately and together using running-sum-class intervals indicate different levels of surfaces but this technique does not reveal the fact as to whether these are structural and tectonic surfaces or are erosion surfaces. The nature of surfaces (whether structural or erosional) may be determined by field observation and ground checks. The altimetric frequency histogram and curve of the Belan basin (U.P.) reveal three frequency maxima viz. (i) at the height of 1000-1200 ft (305 m - 366 m) (X 1), (ii) at the height of 800 feet (240 m) (X 2) and (iii) at the height of 500 feet (150 m) (X 3). Besides, a fourth (though weak) frequency peak is observed at the height of 1350-1400 ft. (427 m). These surfaces (from higher to lower elevations) are, in fact, Kaimur surface (427 m), Parma surface (1000 - 1200 ft or 305 m - 366 m), Rewa surface (800 ft or 240 m) and Trans-Ganga-Yamuna surface (500 feet or 150 m). The crowding of several serial profiles at the same level denotes a surface, the nature of which (i.e. whether of structural or erosional origin) may be determined on the basis of minute field observations and field checks.

Affecting Factors

The nature of modification of erosion surfaces depends on a number of factors viz. age of the surfaces, thickness of later deposits on the surfaces, relative hardness of rocks of which the surfaces have been formed, number of spacing of streams (i.e. stream frequency and drainage density) etc. It is simple assumption that, older the erosion surfaces, greater the modifications and changes i.e. older erosion surfaces are subjected to more changes and modifications by environmental factors than younger surfaces. The changes and modifications of old surfaces may be negated or minimized only when these are protected by overlying thick layers of later deposition of sediments brought down by denudational processes. The degree of changes and modifications suffered by erosion surface depends on the relative resistance of rocks forming those surfaces. For example, if the erosion surface consists of resistant rocks, then after its upliftment relatively longer period of time would be required for the completion of next cycle of erosion because of slow rate of denudation, while the erosion surface formed of weak rocks, when uplifted, would be peneplained in relatively short period of time due to faster rate of denudation. Based on this basic premise it may be inferred that if two erosion surfaces are found side by side at the same elevation but their lithological characteristics are different i.e. one surface has developed on resistant rocks while the other has developed on relatively soft and weak rocks, then the former erosion surface would be much older than the latter surface. The erosion surfaces having high stream frequency and high drainage density suffer more modifications and are destroyed much earlier than the surfaces having poor stream frequency and low drainage density.

There is variation in the nature and form of erosion surfaces. The relatively younger surfaces (i.e. late Tertiary and pre-Quaternary surfaces) and formed on resistant rocks are found in the form of extensive plateaus having accordant heights and flat-topped interfluves. This is because of the fact that such surfaces are least dissected by streams due to lack of requisite period of time. On the other hand, if the surfaces are of pre-Tertiary or Mesozoic age, these are found in much dissected condition wherein the surfaces are segmented into numerous parts which exist in the form of accordant summit levels. If the surfaces are older than Mesozoic, then there is every possibility that the surface might have been obliterated by the dynamic wheels of denudational processes.

Dating System

The correlation and dating (to determine age) of different erosion surfaces developed during different geological periods having varying environmental conditions is difficult geomorphic problem. The most prevalent and widely used method of dating of a particular erosion surface is the correlation of the concerned surface with such erosion surface, developed at other place, the age of which is already known. The height correlation, a method applied for dating purpose, involves the inclusion and consideration of all the remnants of the surface having accordant summit levels. It is believed that the entire erosion surface at the time of its formation might have been of uniform elevation everywhere and might have been continuous but at later dates the extensive surface might have been segmented into numerous parts due to selective but vigorous dissection by rejuvenated streams. It may be pointed out that this method (height correlation) should be applied with caution and utmost care because there is possibility that the original surface might have been deformed due to tectonic movements mainly warping and tilting. The height correlation method yields convincing result in the identification and subjective dating of erosion surfaces of the Chhotanagpur highlands and Ranchi plateau. For example, the Ranchi plateau was peneplained during Jurassic period (prior to the Cretaceous lava flow), the western part of which received basaltic lava during Cretaceous period and was subsequently uplifted during Tertiary epoch and is named as 'patland' (3500 feet or 1065 m a.m. s.l.). The patland has been segmented and fragmented by the North Koel and its tributaries into numerous mesas and buttes, locally known as 'pats' (e.g. Netarhat pat, Jamira pat, Khamarpat, Raldami pat, Rudni pat, Bagru pat, Lota pat etc.) having accordant height. Inspite of this, height correlation method is generally used for dating the post-Tertiary and Quaternary (if any) erosion surfaces and not for the pre-Tertiary surfaces.

Erosion surfaces are also tentatively dated on the basis of geological unconformity. It generally happens that the erosion surfaces, after their formation, receive sediments and are, thus, buried under thick cover of sediments and remain protected from environmental processes so long as surficial cover is not removed. Some times the surficial covers of sediments are removed, the buried surface is resurrected and is exposed to environmental conditions. Consequently, the exposed erosion surface is subjected to changes and modifications by denudational processes. In such cases, the dating of erosion surface becomes difficult but wherever the erosion surface is well protected under surficial deposits, the dating of the surface becomes easy proposition. This method is also not free from pitfalls because even the surficial deposition may also experience changes and modifications due to weathering processes, mixing of older deposits with newer deposits etc. It may also be pointed out that the geomorphic processes involved in the formation of a particular surface are not always the result of climate. For example, if the surficial sediments of marine origin are found over an erosion surface, it is not necessary that the underlying erosion surface is the result of marine process. It may be possible that the concerned surface might have been formed by fluvial process close to the sea coast and the marine sediments might have been deposited on such surface in two ways viz. (i) marine sediments resting over nearby high marine surface (coast land) might have reached fluvially originated erosion surface through soil creep and slumping or (ii) the rise in sea-level might have caused transgression of sea water upto fluvially originated erosion surface which might have facilitated deposition of marine sediments on it. The subsequent regressional phase of sea caused by negative change (fall) in sea-level might have caused retreat of sea water and thus the erosion surface having marine sediments might have been freed from sea water and exposed to environmental conditions. The development of consequent streams on such surface might have deposited fluvial sediments over marine sediments and thus the mixing of marine and fluvial sediments resting over an erosion surface complicates the process of dating of such erosion surfaces. Thus, the dating of erosion surfaces on the basis of sediments resting over them should be done with great care and precaution. According to L.C. King the description of sediments resting over an erosion surface gives clue only for the end date of the formation of the surface by final processes. The entire process of the formation of the surface is not revealed by this method. According to him the Gondwana pediplain surface in South Africa was formed prior to the breaking of Gondwanaland but these surfaces have been wrongly dated as Tertiary because these carry the deposits of Miocene period.

Another method of dating of erosion surfaces and determining of mechanism of the formation of erosion surfaces involves the interpretation of sequence of sedimentation in another area but close to the concerned surface. It generally happens that the eroded materials of highland are deposited in the nearby lowlying areas. The sequence of deposition of sediments from below (in the beginning) upward involves conglomerates having boulders, cobbles and pebbles followed by medium-sized particles in the middle and fine grained clay and marl in the uppermost part. On the basis of such cycle of deposition the successive stages of erosion of highland, subsequent slope decline and development of erosion surface are understood.

The determination of the mode of origin of erosion surfaces and their dating are also accomplished on the basis of drainage patterns developed on those surfaces. This involves the reconstruction of the evolutionary history of drainage of the concerned region.

Radio carbon-dating is another very useful method of determining the absolute age of erosion surfaces but this method cannot be applied for those surfaces which were formed prior to the evolution of organic life or where organic fossils have been completely destroyed.

Various Erosion Surfaces

Chhotanagpur Highlands : The present description is entirely based (rather is adapted) on the classical work of Prof. R.P. Singh (Geomorphological Evolution of Chhotanagpur High lands, 1969, NGSI, Varanasi, pp. 108-120), who has identified the following erosion surfaces on the basis of interpretation of past cycles of erosion (complete and incomplete), geological structure, deposits of sediments, drainage system and patterns, morphometric techniques, field observations etc.

Pre-Dalma Surface : The geomorphic history of Chhotanagpur begins from the upliftment of the region in Archaean period which resulted in the formation of folded structure of Chhotanagpur region and adjoining areas. Initially, trellis drainage pattern developed in conformity with the folded structure. The uplifted landmass was eroded down to base level and thus was peneplained by fluvial cycle of erosion known as pre-Dalma cycle of erosion but this cycle was interrupted due to lava flow. Though no perceptible evidences of upliftment during lava flow are available but folded lava sheets in Porhat hills speak of the upliftment even after lava flow. It is very difficult to trace and identify pre-Dalma erosion surface because several cycles of erosion have succeeded pre-Dalma cycle and hence the evidences of initial (first) cycle and erosion surface have been completely obliterated due to polycyclic reliefs. At some places the Dalma lava sheet has succeeded in preserving the traces of initial erosion surface e.g. a few erosional unconformities are observable in Dhanjori highlands where pre-Dalma surface has been overlain by Dhanjori sandstone conglomerate.

Pre-Cambrian Erosion Surface : Fluvial processes were dominant from Dalma lava flow to pre-Cambrian period and thus granitegneiss base was exposed due to continued fluvial erosion. Consequently, dissected gneiss surface was formed. Erosional unconformities are again found in Kolhan highlands but these are younger than Dhanjori unconformity because 'Kolhan sandstone-conglomerates' have been deposited horizontally on dissected granite surface. Similarly, 'Vindhyan sandstones and limestones' have been deposited on 'pre-Cambrian granite surface' in the north-western part of the Chhotanagpur region.

Carboniferous Erosion Surface : Gondwanaland was extensively and intensively glaciated during Carboniferous period, the evidences of which are found in the form of glacial boulders in the coal seems of Talchir coal mines of Orissa (in the case of Chhotanagpur highlands). The pre-existing cycle of erosion was interrupted (rather the chapter of cycle of erosion was closed) due to Carboniferous glaciation. The ice sheets covered the existing surfaces and rounded the hills and peaks as representatives of earlier cycles. Thus, the pre-existing topography was so greatly changed and modified by advancing ice sheets that all the evidences of earlier cycles and landscapes were destroyed. Thus, a monotonous topography might have been formed due to Carboniferous glaciation. There might have been absence of well developed and established drainage system on this surface. Marshes, swamps and lakes might have been formed by retreating ice sheets during interglacial periods. The gradual integration of streams might have given birth to the evolution and development of well developed drainage system.

Permian-Triassic Erosion Surface : A new chapter of denudation and landscape development was added to the geomorphic history of the Chhotanagpur region after the retreat of Carboniferous ice sheets. The highlands were greatly denuded. The area was reuplifted during Triassic period and the Gondwana formation was rapidly eroded. Some sort of tectonic movements resulted in the formation of vertical faults but the horizontal character of sediments could be maintained and preserved. The region was peneplained by Triassic period.

Gondwana Surface : Permian-Triassic cycle of erosion was interrupted during late Jurassic and early Cretaceous periods due to volcanic eruptions and widespread lava flows. Peninsular India was separated from Gondwanaland during Cretaceous period. Rapid fluvial erosion started due to origin of monsoon climate. This fluvial cycle of erosion is called as Gondwana cycle and the surface produced by this cycle is called Gondwana surface. The identical and contemporary Gondwana surfaces are also found in other parts of Gondwanaland viz. South America, Africa, Peninsular India, Australia etc. This cycle is also called as pre-tertiary cycle which was terminated by Deccan trap in India. Most of the drainage lines (streams) were obliterated due to lava cover. The Gondwana surface was extensive, undulating and low in elevation and sloped eastward i.e. towards pre-tertiary sea. It is estimated that the elevation in the east and west might have been 305 m and 488 m respectively. Some sort of upliftment of this surface cannot be ruled out. This resulted in breaks in slopes at certain places.

(6) Tertiary period registered three phases of upliftment as a consequence of Himalayan orogeny which effected upliftment of existing surface, tilting, faulting, subsidence etc. The first upliftment occurred during Miocene period when Chhotanagpur was uplifted by 305 m. Consequently, the streams were rejuvenated and new cycle was initiated. Second phase of upliftment occurred during Pliocene period when the western part of Chhotanagpur was uplifted by another 305 m. The third phase of upliftment was upto 214 m to 305 m. The total upliftment of south-western Chhotanagpur was 915 m (3000 ft). The highly dissected margins of Ranchi plateau, Hazaribagh plateau, Kolhan - Baghmundi Highlands reveal Tertiary upliftment of the region. Breaks in longitudinal profiles of the Subarnarekha (Hundrughaugh falls), the Kanchi (Dassamghaugh falls), the Raru (Gautamdhara or Johna falls) streams etc. indicate upliftment of Ranchi plateau and rejuvenation of these streams. Breaks in longitudinal profiles of upper reaches of the North Koel, the Sankh and the Burha rivers validate upliftment of western Ranchi plateau and Palamau upland during late Tertiary epoch. The transverse sections of major streams of the Chhotanagpur highlands reveal marked incision of valleys due to rejuvenation.

Tertiary cycle of erosion began with the upliftment of the western Chhotanagpur. This cycle was interrupted because of mid-tertiary upliftment and thus mid-tertiary cycle, also known as Hundru falls cycle (R.P. Singh) was initiated. Juvenile stages of streams along the margins of plateaus present significant evidences of this cycle. Numerous waterfalls are found in N.E.-S.W. direction. These falls are characterized by steep scarps, truncated strata, deep and narrow gorges, etc. Tertiary cycle of erosion was also interrupted in the S.E. Chhotanagpur due to moderate nature of upliftment.

The present landscape of Chhotanagpur region is polycyclic in origin and presents and ideal example of palimpsest topography.

Peninsular Indian Denudation Chronology : The present description on denudation chronology of Peninsular India has been based (rather adapted) on scholarly works of R.P. Singh e.g. 'Geomorphological Evolution of Chhotanagpur Highlands, 1969 and 'Landscape Cycles of Peninsular India, 1967 and reprinted in 1983.

Present Physical Set Up : The present physical set up of Peninsular India is the result of different phases of upliftment, denudation, lava flow etc. The present landscape mosaic is characterized by polycyclic reliefs. Peninsular India is bordered by old and highly dissected relict mountains and hills. Aravalli ranges extend from Gujarat to Delhi in the north-west and represent oldest crustal part of the country and inspite of high degree of denudation through ages still have such peaks which are 1220 m to 1525 m (4000 to 5000 ft) at mean sea-level wherein Mt. Abu is the highest peak (1693 m or 5650 feet a.m.s.l.). A few streams cut the Aravallis across and drain towards Arabian Sea (e.g. Mahi, Luni) and a few meet the Yamuna (e.g. Chambal and Banas). Malwa plateau is surrounded by Vindhyan ranges. In fact, Vindhyan ranges appear as escarpments of rift valleys. Vindhyan ranges running from Indore to Bhopal, through Baghelkhand extend upto Sasaram in Bihar. Kaimur ranges form escarpments almost parallel to the Son valley. Malwa plateau has been so greatly dissected by the streams that the whole area appears as ravinous region. Bundelkhand region has been most severely ravinated and Chambal ravines are famous world over. The Deccan tableland extends in triangular tabular form to the south of the Narmada valley. This triangular tableland is bordered by hills and ranges on all of its three sides. Satpura hills extend from Rajpipla in the west to Rajmahal hills in the east. The domed structure of Satpura is buried under Deccan lava in the west while in the east Mahadeo hills (Pachmarhi hills) of Gondwana system become prominent reliefs. Maikal hills composed of Archaean shield and Deccan traps are significant reliefs near Amarkantak. These are further extended eastward in the form of Surguza, Ranchi and Hazaribagh ranges. The most outstanding characteristic feature of the peaks of the hill ranges of Peninsular India is that they have flat tops rather than sharp edges. Satpuras have been affected by faulting activity to the largest extent and thus the streams flow through deep and narrow gorges. The valleys of the Narmada and the Tapi rivers to the north and south of Satpuras respectively are fine examples of rift valleys. It may be mentioned that the Narmada and Tapi rivers flow opposite to the general slope of Deccan plateau (i.e. westward).

Western Ghats extend from the estuary of the Tapi river in the north to the Cape Comorin in the South. The stepped features of the western face of the Western Ghats indicate evidence of faulting activity. Very few streams cut across the Western Ghats. The Eastern Ghats extend in discontinuous manner from Mayurbhanj in the north to Nilgiri in the south. The eastward flowing major streams (e.g. Mahanadi, Godavari, Krishna, Penner, Kaveri etc.) have dissected and segmented Eastern Ghats. Nilgiris join Eastern

Ghats with Western Ghats. Palghat separates Annamalai hills from Nilgiris. The northeastern segment of Annamalai hills is known as Palni hills. In the north-east, the hills of Peninsular India are known as Assam hills.

The Indian peninsula has undergone different phases of extensive faulting, tilting and folding activities in geological past. Faulting has been responsible for the segmentation of Peninsular India into several small plateaus and tablelands e.g. Malwa plateau, Deccan tableland, Mysore plateau, Chhotanagpur plateau etc. Tributaries of the Damodar, the Subarnarekha and the N. Koel rivers have further segmented Chhotanagpur plateau into still smaller plateaus. The Peninsular India has experienced several cycles of erosion and hence the present topographic features are representatives of polycyclic reliefs.

Landscape Evolution : The present landscape mosaics of Peninsular India have evolved through different phases of numerous cycles of erosion, sedimentation, lava flows, orogenesis, metamorphism, faulting, tilting, eustatism, resurrection of palaeo landforms etc. and thus the present landscape is an example of palimpsest topography. About half of the oldest structures has been exposed due to prolonged denudation of super incumbent formations and covers. If we look into the cyclic development of landscape of Peninsular India, it appears that numerous evidences of positive and negative changes in sea-level (and hence in base level of erosion), lava flows, climatic changes, interruptions in cycles of erosion etc. are clearly observable. 'While the imprint of some cycles has been preserved and can be deciphered, that others have been obliterated by the ceaseless march of time' (R.P. Singh, 1960).

Geological Structure : Geological structure plays an important role in the evolution of landscapes. Though, some of the initial structures of Peninsular India have disappeared due to prolonged denudation, yet the existing structure is competent enough to speak of its morphological features. The geology and structural history of Peninsular India is related to the rigid masses of Gondwanaland. A brief description of stratigraphic characteristics of Peninsular India is necessary for the reconstruction and interpretation of its denudation chronology. The geological history of the region began with the formation of Archaean group of rocks and terminated with the deposition of recent sediments of Aryan group (recent delta deposits, recent raised beaches, coral coast etc.). The process of sedimentation still continues.

(1) Archaean group of rocks consists of Archaean granites and genesis and Dharwarian sedimentaries. Dharwar system of rocks includes Dharwar sedimentaries of Dharwar and Bellary districts of Karnataka, Sausar series of Karnataka and Chhotanagpur, Chipli series of Balaghat and Chindwara districts of M.P., Gondite series of Rewa and Jabalpur districts of M.P., Kodurite series of Vishakhapatnam district of Andhra Pradesh, Dalma traps in Singhbhum district of Bihar and

Aravalli series. The Dharwar formations have suffered maximum metamorphism and lack in fossils. Rocks are not found in their original horizontal manner, rather these have been greatly deformed by tectonic activities (folding, fauting, tilting etc.).

(2) Pruana group of rocks comprises two major categories i.e. (a) Cuddapah system of rocks, and (b) Vindhyan system of rocks. Rocks of Cuddapah system were deposited from Cambrian period to pre-Algonikan period and these include shale, slate, quartzite and limestone. The rocks have been greatly metamorphosed and hence lack in fossils. The Cuddaph formations are divided into two groups viz. upper and lower Cuddapah. These rocks are found in the form of Krishna series in Krishna valley, Nallamalai series in Nallamalai ranges, Cheyar series in Cheyar valley, Papaghani series in Papaghani valley, Delhi series in Delhi hills etc. Vindhyan formations include an extensive area of 1,00,000 Km^2 which extends from Sasaram of Bihar in the east to Chittorgarh in the west. Vindhyan formations include sedimentary rocks e.g. sandstones, shales, limestones etc. These are further divided into lower and upper Vindhyans wherein Semri series (in Son valley), Karnool series (Andhra Pradesh), Bhima series (Bhima valley), and Palni series (Jodhpur and Chittorgarh of Rajasthan). Some chunks of Vindhyans are also found in Malwa plateau and Bundelkhand.

(3) Dravidian group consists of several systems of rocks of which upper Vindhyans are most significant. These were deposited in Cambrian period in river valleys. The major constituents are sandstones and conglomerates. These include Bhander series, Rewa series and Kaimur series mostly in M.P. and Rohtas plateau of Bihar. The rock beds have almost preserved their horizontal character.

(4) Aryan group — The rocks of Aryan group were formed from Permian to Jurassic periods. Gondwana system of rocks is the most significant constituent of this group. Most of the Gondwana rocks were deposited in river valleys and have been least disturbed and deformed and hence they almost maintain their original horizontal disposition. Gondwana formations contain most of Indian coals (98 per cent). With reference to depth Gondwana formation is divided into 3 categories viz. (a) lower Gondwana system (Permian to Carboniferous), (b) middle Gondwana system (Triassic period), and (c) upper Gondwana system (Jurassic period). Gondwana formation is widely found in Godavari valley, Mahanadi valley, Venganga and Wardha (tributaries of Godavari) valleys, Kutchch and Kathiawar of Gujarat, western Rajasthan, Damodar Valley, Vishakhapatnam etc. Lower Gondwanas are found in the form of Talchir series, Damodar series and Raniganj series which have very rich deposit of coal. The middle Gondwanas include Maleri sub-series, Mahadeo sub-series and Panchet sub-series.

There was an extensive lava flow during Cretaceous period which covered an area of 2,00,000 square miles (5,18,000 km^2) of Peninsular India. The depth of lava sheets ranges between 2000 feet (610 m) to 5000 ft (1520 m), the greatest thickness being 10,000 feet (3050 m). Deccan lava traps are found in Gujarat, Maharashtra, Saurastra (Gujarat) and western M.P. but some patches of trappean lava are also found in Bihar. Deccan traps are divided into (i) lower traps (5000 ft or 1520 m thick), (ii) middle traps (4000 ft or 1220 m), and (iii) upper traps (1500 ft or 457 m).

The Tertiary system of rocks includes clay, gravels, sandstones and limestones of Eocene period in Gujarat; sandy limestones of Oligocene to Pliocene in Kathiawar and the *extreme eastern* and western margins of the peninsula ; (from lower to upper) argillaceous sedimentaries, sandstone - limestones, clays, marl and calcareous rocks in Kutch area; limestones of Eocene period in Bikaner, Jodhpur and Jaisalmer of Rajasthan; fossiliferous deposits from Eocene to Pliocene periods on entire Coromondal coast; limestones having fossils of molluscs, corals, foraminifera etc. on Malabar coast etc.

The Pliocene deposits include Pliocene laterites, upper Cuddalore sandstones, sand dunes of Rajputana, loess, raised beaches on sea coasts, Porbandar sandstones, gravels of *pre-Palaeolithic* age, *low-level laterites*, older alluvia of Narmada and Godavari valleys etc.

If we look into the geological history of the Peninsular India, it appears that no evidences of large-scale submergence under sea and orogenesis are found. Evidences of only local transgression of sea are found when Cuddapah and Vindhyan sedimentaries are supposed to have been formed under marine environment. Later transgression, if any, involved only coastal areas. Tearing events in the valleys of the Son, the Damodar and the Mahanadi rivers might have occurred due to Hercynian orogenesis. The foreland of Indian peninsula was indirectly affected by the side effects of the Himalayan orogeny during Tertiary epoch, resulting in localized upliftment, upwarping, downwarping etc.

The denuded outcrops of the Peninsular India reveal the fact that initial orogenesis resulted in the formation of westward tilted folds followed by intense folding resulting in the formation of north-northwest oriented anticlinorium. The formation of Delhi - Arawalli ranges became responsible for the development of N.E.S.W. trending synclinorium. Archaean mountains are supposed to have been peneplained before Dalma and Dhanjori lava flows. According to R.P. Singh the landscape development of the Peninsular India has passed through the following stages - (present description is adapted version of R.P. Singh's research paper entitled, 'landscape cycles of Peninsular India)'.

Developmental Stages

Pre-Dharwar Landscape : The geomorphic history of Peninsular India began with the primeval original solid crustal surface which suffered the activities of denudation and sedimentation for longer period of time. These sediments (known as pre-Dharwarian sedimentaries) were buckled, folded and metamorphosed several times and thus basal granites and genesis of Peninsular India were formed. There was intrusion of magma in these rocks at certain localities, the evidences of which are still observable in the form of charnockites of Nilgiri, Palni and Shevroy. It is surmised that five ancient geosynclines might have existed before Cambrian period e.g. Dharwar geosyncline (which extended from present Dharwar to Mysore), Eastern Ghat geosyncline, Satpura geosyncline, Aravalli geosyncline, and Delhi geosyncline.

Dharwar Landscape Cycle : The Dharwar erosion cycle began with the formation of initial mountains out of the aforesaid geosynclines. These mountains are supposed to have been peneplained by the dynamic wheels of fluvial erosion before Cambrian period, though according to Heron Aravallis are considered to be peneplained by Mesozoic era. The activities of erosion, sedimentation and magmatic intrusions and lava flows continued for longer period of time. Numerous longitudinal consequent and some transverse streams were formed with Arachaean folding. Evidences of sedimentation (Dhanjori conglomerates) and igneous activity (Dalma and Dhanjori lava flows) have been found in these valleys. Archaean lava flows played significant role in controlling the landscape development right from Madhya Pradesh, through Chhotanagpur to Medinipur of West Bengal and unintegrated drainage system was evolved.

Cambrian Landscape : Cycle was initiated with the ablation and obliteration of ice sheets from Vindhyan ranges, which continued for longer period of time covering Vindhyan formations resulting in substantial changes in reliefs and drainage. Aravallis were lowered in height due to intensive denudation but the Aravallis could be peneplained only by the end of Jurassic period and beginning of Cretaceous period. Though there was significant reduction in reliefs of Peninsular India due to Cambrian cycle of erosion, but still recognisible reliefs could be preserved.

Carboniferous Glaciation and Landscape : The entire Gondwanaland (including S. America, Antarctica, Africa, Malagasi, Peninsular India, and Australia) was extensively glaciated during Carboniferous period. Though evidences of Carboniferous glaciation have been mostly destroyed due to upliftment and subsequent denudation, yet some scattered relics of this glaciation enable us to suppose that the axis of the glaciation might have been over Aravallis from where the ice sheets might have spread in eastern and south-eastern directions along the axis of Aravalli - Hazaribagh ranges. Sufficient evidences have been found to demonstrate the extent of Carboniferous glaciation from Rajmahal to Godavari and from Raniganj to Nagpur. There were fluctuations in Carboniferous sea-levels due to advancement and retreat of ice sheets which resulted in transgression of sea on land. The previous cycle of erosion was terminated because of spread of ice sheets on existing topography. The lower reaches of a few streams became truncated because these could not be covered with advancing ice sheets. Talchir basin is supposed to be in limnological stage, as a result of which boulders were deposited upto the depth of 15 to 30 m (50 - 100 ft). In fact, lakes were formed due to initial moraines which (lakes) enabled deposition of boulders at a later date, the limnological stage of Talchir basin was terminated because of on set of warm condition which facilitated the deposition of rocks, rich in iron content. Raniganj and Panchet ranges are the result of such formation. The Hercynian movement caused tearing of peninsular gneissic surface which resulted in the formation of several tectonic troughs. Such activities were more frequent in the basins of the Mahanadi, the Damodar and the Godavari rivers wherein sinking valleys allowed sedimentation upto thousands of feet in thickness. It may be mentioned that the process of sedimentation in these valleys was accomplished by ,melt-water streams' (due to deglaciation of ice sheets).

Gondwana Landscape Cycle : The Carboniferous glaciated surface is separated from subsequent Gondwana sedimentaries by an unconformity. Gondwana sediments are still preserved in tectonic basins, though these have been greatly modified. It may be mentioned that Rajmahal basalt (due to lava flow) was also formed during the phase of Gondwana sedimentation. The peninsula was peneplained by the end of Mesozoic era i.e. by early Cretaceous period. The areas not affected by intensive denudation were characterized by the formation of depositional landforms (aggradational landforms). It is interesting to note that the general slope of Peninsular India was directed northward during Gondwana cycle of erosion but later on this slope axis changed due to tilting. Though most of the Gondwana erosion surfaces have been destroyed due to subsequent erosion but their traces are still found in Tamil Nadu, Madhya Pradesh and Maharashtra. R.P. Singh has opined that relics of Gondwana surface may also be found on Aravallis and Vindhyan ranges.

Fragmentation of Gondwanaland and Post-Gondwana Landscape Cycle : Gondwanaland was fragmented in mid-Mesozoic era and Peninsular India was tagged with Asia. The marginal tracts of the peninsula were transgressed by sea several times resulting in the formation of horizontal structure of lemeta formation. Kutch received marine sediments. New landforms were added due to erosion of Gondwana cycle during post-Gondwana cycle in early Cretaceous period. It is believed that this cycle of erosion was completed, the evidences of which are still preserved in the form of bevelled surfaces of Udagamandlam (Ooty) and Cordmum ranges, high summits of the Eastern Ghats, relics in Maharashtra and Madhya Pradesh and highlands of Vindhyan ranges.

Cretaceous - Eocene Lava Flow : There was lava flow in the north - eastern section of the Peninsula i.e. in Rajmahal area during Cretaceous period covering an area of 3,96,917 km^2 and pre-existing topographic features were extensively modified as the previous topography was buried under lava sheets. There was further extensive lava flow during Eocene period which covered the denuded Archaean, Vindhyan and Lemeta surfaces. Thus, Deccan lava flow covered an area of 5,18,000 km^2 extending from Kathiawar to Nagpur and from Malwa to Dharwar. This lava formation is called as Deccan trap which was subjected to severe weathering and erosional processes for fairly a long period under monsoon climatic conditions. The basaltic lava was extensively weathered to form laterites, the well developed profiles of which are observable over Chhotanagpur and Mahabaleshwar plateaus. The prolonged denudation of basaltic surface by fluvial process resulted into the development of hills, ridges, valleys, plains etc. A few of these hills are still observed. It may be mentioned that these hill ranges have been formed because of denudation and not due to folding and upliftment. In the region of Vindhyan formations basaltic cover has been removed due to denudation at certain places and hence older Vindhyan topographic features have been uncovered. The scarps of basaltic plateau are still undergoing the process of parallel retreat due to backwasting. Wherever the parallel retreat is over and the lateritic cover has been removed, the hills have been modified to conical shape having almost convexo-concave slope plan. Such hills presently constitute significant landscape components of Peninsular India.

Cenozoic Landscape Cycle : The topographic reliefs of Peninsular India were peneplained before Tertiary epoch. The western margins of Western Ghats of India came into existence due to subsidence of extensive area between Africa and Indian peninsula and subsequent evolution of the Arabian Sea. Western Ghats determine the drainage axis of present drainage system but the Narmada and the Tapi rivers are not conformal to this drainage axis as they flow westward. In fact, these rivers flow through rift valleys which have been formed due to side effects of the Himalayan orogeny. Vindhyan and Kaimur ranges to the north of the Narmada - Son troughs, Satpura and Mahadeo ranges to the south of the Narmada valley, and Ajanta ranges to the south of Tapi valley are in the form of scarps. Upwarping at a later date in north-south direction caused breaks in long profiles of the Narmada, the Tapi and the Godavari rivers. The evidences of breaks in the long profiles of rivers caused by tilting of Indian peninsula are presently found in the form of nick points (water falls). In fact, some of the rivers of the peninsula were rejuvenated due to Tertiary uplifts, tilting and warping which is evidenced by the presence of multi-level plateaus, highlands, nick points followed by waterfalls and gorges. The presence of raised beaches (30 m - 45 m) validates negative change in sea-level (fall in base level of erosion) and consequent rejuvenation of streams.

Such topographic feature is considered to be the result of Cenozoic erosion cycle which is characterized by accordant summit levels. Mysore plateau, Nilgiri, Cordmum-Annamalai ranges are without doubt the examples of Cenozoic peneplains which are characterized by relict denuded hills representing Davisian monadnocks. The high - level surfaces of Khondalite and charnokite rocks in Andhra Pradesh and Orissa represent highest erosion surface. The gneissic surface of Bastar range (610 m-915 m) represents second erosion surface while dissected domes of Keonjhar, Mayurbhanj and Sundargarh districts (610 m) of Orissa represent third erosion surface.

Extensive warping occurred during Quaternary period which rejuvenated most of the rivers resulting in the development of numerous nick points, waterfalls and gorges in their longitudinal profiles. The rejuvenation initiated valley deepening and headward erosion. Significant resultant waterfalls include Sivasamundram falls (300 ft or 91 m), Goka falls (180 feet or 55 m), Gyrsoppa falls (850 feet or 259 m), Dhunwadhar falls (30 ft or 9.14 m) etc.

There was large-scale changes and modifications in the existing drainage system. Khambat (Combay) Ahmedabad-Son traces of transgressional phase of sea are found on the most severely denuded surfaces (i.e. Cuddapah and Vindhyan Seas). Cuddapah sea is supposed to have extended upto Nilgiri hills and over certain parts of Madhya Pradesh. The intervening areas between Vindhyan and Cuddapah seas are devoid of marine fossils. The Vindhyan sea might have extended from Aravallis in the west to Sasaram in Bihar in the east. The Narmada and Son valleys are supposed to have been covered by Bizawar sea but this might have been obliterated before Cuddapah sedimentation. Vindhyan sea was further extended due to subsidence of surface to the north-west of Bundelkhand facilitating deposition of Gwalior system as observed by Oldham.

Cuddapah : Vindhyan Landscape Cycle was initiated with the upliftment of Cuddapah and Vindhyan formations. Streams coming out of the Western Ghats and southern slopes of Satpuras started to dump sediments, brought down by them, into Cuddapah Sea. Similarly, the streams taking their sources from the Aravallis and northern slopes of Satpuras began to deposit sediments into Vindhyan Sea. Thus, there existed three drainage axes of Arachaean drainage viz. Western Ghat drainage axis, Satpura drainage axis and Aravalli drainage axis. Numerous consequent, subsequent and obsequent streams developed with the origin of Vindhyan ranges. The evolution of this new drainage system resulted in the readjustment of existing Aravalli and Satpura drainage systems.

Vindhyan Glaciation : The existence of boulder bed of the Son valley, tillites of the Ken Valley, Pandava falls and 'fluvial constinental de posits' of Semri series validates the corollary of glaciation of the Vindhyan ranges. It is supposed that the upper parts of Vindhyan ranges might have been covered with glaciers and hence the sources of streams might have been obliterated. It is further believed that the surrounding areas of these glaciers might have been characterized by periglacial processes. But the truncated streams (as their sources were covered with ice) might have become successful in maintaining their middle and lower courses. Glaciers were soon ablated, region received marine deposit over its gneissic peneplain surface, lava surface and other older rocks due to transgression of sea during Pleistocene period. Gujarat experienced recent upliftment while emergence of land occurred in Khambat (Combay) area. The present coastal tracts are characterized by raised beaches, sand dunes, lagoons and alluvial deposits.

Denudation Chronology of Belan Basin : The present description of denudation chronology and erosion surfaces of the Belan basin representing Vindhyan formation of sedimentaries and Quaternary sedimentation is reproduced from the research paper entitled, 'Geomorphological evolution and erosion surfaces of the Belan basin' published by Savindra Singh, and Renu Srivastava, 1976 (National Geographical Journal of India, Vol. 22, Nos. 3-4, pp. 124-138).

Physiographic Personality : The Belan basin, occupying a synclinal trough probably of upper Vindhyan age between the Vindhyan scarps of the northern bank of the Son river (Kaimur range) in the south and water divide of the Gangetic tributaries and northern offshoots of its own in the north (24°35'N — 25°2'30" N and 81°45' E- 83°15 E) is bordered by Chhotanagpur plateau in the S.E. and the Tons valley in the west. The basal rocks are sandstones and shales of Kaimur, Rewa and Bhander series of upper Vindhyans. Overlying Quaternary deposits include (from lower to upper) gravel - mottled clay formation, red-brown graveland formation, yellow silt formation, caliche formation, buried soil layer, aeolian sand formation, and modern soil horizon.

The present relief features of the entire Belan basin resemble the usual landscape of the hilly tract and are marked with subdued relief. The southern part is characterized by Vindhyan scarpland overlooking the Son valley in the south, acting as an effective water divide between the Belan and the Son systems. It is a striking feature that no tributary except Ghaghar and seasonal torrents descend the scarp to meet the Son from the south but the Belan receives a number of tributaries from the northern face of the scarp. In the extreme south-eastern part of the basin, Bijaigarh highlands having entrenched valleys have rare similarity with the general relief of the basin. Jagged between Vindhyan scarps (Kaimur hills) in the south and Mirzapur hills in the north the Belan river has cut steep banks rising to considerable heights which in some cases extend upto 18 m-20 m. The hills bordering the basin are plateau - like at their crests and slopes have been carved out in a series of steps guided by prominent joint planes in the rock formation. The Belan river and its tributaries have formed rectangular drainage pattern wherein most of the streams join the Belan river at approximately right angle guided by the joints of rocks. The Adwa river, a tributary stream of the Belan river, cuts through an almost peneplained valley with slight undulation. Occasionally small rounded to ellipsoidal mounds prominently stand to form the peneplained terrain, although to the west of the Adwa valley high mesas and buttes carved out of the Rewa series of the Upper Vindhyan formations, form conspicuous landmark topography of the area. The Belan-Seotilocality is also characterized by the presence of several mesas and buttes representing the relics of erosion surfaces. Numerous falls are located on the Belan and its tributaries which have been originated by frequent faulting by tectonic impulses.

Geological History : The geological history of the basin begins with the sedimentation of Vindhyan Sea during Algonikan era, when upper Vindhyan sandstones and shales were deposited. If the filling of Vindhyan sea-part of the Belan basin is taken to be complete upto Permian period, the region must have experienced several tectonic and gradational processes which resulted into the evolution of present topography of the basin. 'Whatsoever might have been the geography of the area during its stage of initiation, the present surface morphology with respect to drainage, slope, relief etc. bears adequate relationship with the regional and local faults, which are numerous in the region' (R. L. Singh, 1974). Before the initiation of faulting in the Belan basin in particular and eastern Vindhyan upland in general, two main altitudinal zones, as observed by R.L. Singh, might have existed viz. (i) valley of the Son somewhere at the elevation of 1000 ft (305 m) above sea-level and (ii) highlands on both the banks of the Son rising to about 2500 ft (762 m) above sea-level. The region south of the Belan basin seems to have experienced two episodes of reverse faulting running in east west direction. This faulting was responsible for northward shifting of the Son valley and also the tilting of the northern block to the north which is proved from the fact that, 'the Vindhyan sandstones which are exposed in the Son gorge above 1000 ft (305 m) from sea-level rising upto 1500 ft. (457 m) are seen between 750-1000 ft (229-305m) in the Belan valley and buried below the Ganga alluvium with exposures along the river, one of which forms the base of Chunar Fort' (R.L. Singh, 1974).

A series of rapids, deep pools, variation in the attitudes of rock-beds, deflections in the river courses at certain places are the clear cut testimony of a series of faults which occurred during Tertiary period. The displacement of flaggy beds in the rock sequence and the presence of conchoidal fracture clearly indicate that the region might have been displaced along the fault plane.

As regards the erosional history of the Belan basin and environs, the upper Vindhyan rocks were exposed to continuous processes of weathering and erosion but the region could not attain its maturity until Cretaceous period which is indicated by Kaimur erosion surface (R. S. Dubey, 1968) in the southern part of the basin. 'The problem why the Vindhyans were not peneplained even after a long period from Cambrian to Cretaceous could be explained by assuming the possibility of the Vindhyan rocks being protected by the cover of ice for some time' (R. S. Dubey, 1964). Following D.N. Wadia (1961) two glacial periods in the past, one during the upper Carboniferous and other during Bunter age of the lower Triassic period might have acted as covering sheets of the Vindhyan rocks.

The Panna surface of R.S. Dubey (1968) in Rewa Plateau is scanty in the Belan basin owing to the fact that Eocene period might have not been much active in raising the region upto required height. The Tertiary grits and gravels of Miocene period are separated by unconformity from the Upper Vindhyan bedrocks. Later Tertiary earth movement during post-Pliocene period was responsible for slight upheaval which resulted into the formation of Rewa erosion surface in the western part of the basin. The northward tilt of the southern upland south of the Belan basin due to Tertiary tectonics and a subsequent syncline formation resulted into the grading of the Belan river and attaining its present mature stage. The continuous denudational processes during Quaternary epoch ultimately formed trans-Yamuna-Ganga surface at the lowest level.

The stratigraphy of Quaternary alluvial deposits over Vindhyan bed-rocks helps in reconstructing the complete chronology of the climatic events. The lateritic crusts lying over Upper Vindhyans are suggestive of humid warm tropical climate related to upper Pleistocene interglacial period which was followed by an episode of cold glacial climate related to latest Pleistocene glacial period during which gravel-mottled clay was formed over lateritic crust and somewhere over Vindhyan formations directly. The cold climatic phase is also supported by the records of facetted, grooved and striated gravels, cobbles and boulders from the excavation in the Belan valley about 32 km S.S.W. of Mirzapur by Krishnaswami and Hukku. The cold climatic phase later on changed into increased aridity with the result large-scale transportation of big boulders and cobbles was stopped and they were deposited in situ.

After the expiry of cold phase and consequent aridity, relative humidity increased resulting into pronounced rainfall which accelerated large-scale transportation of large-sized boulders and pebbles and consequent decrease in the grain-size during transportation. This humid climate accelerated the dynamic wheels of erosion resulting into intense dissection and the removal of gravel-mottled clay formation making unconformity between Vindhyan formation below and red-gravel-sand formation above. A partial oxidation introduced speckling in clay formation.

The aeolian phase marked by relative aridity was again restored and red gravel-sands were deposited over which developed gullies and arroys at later date during humid phase related to post-glacial period. The wavy nature of boundary between red-gravel-sand formation and subsequent deposits clearly indicates a humid period of pronounced and net erosion and prolonged phase of weathering when the process of 'rubefaction' was in progress (H.L. Movius Jr. cited by Kirk Bryan and C.A. Albritton Jr., 1943). This humid phase could not last long and was subsequently changed to pronounced arid climatic phase again during which a thick envelope of Liassic deposit covered almost all the earlier deposits. The reappearance of relative humidity again was responsible for solution and pitting of the upper horizon of the caliche formation. This humid phase was temporarily intervened by a shorter period of arid phase when numerous Liassic mounds were formed. The present day climate, still experiences dry spell but the runoff is sufficient for gullying and arroying the present day surfaces. The denudation chronology of the Belan basin may be presented in tabular form as follows—

From Youngest to the Oldest

Eras/Epoch	*Period*	*Climate*	*Geological Events*	*Erosion Surfaces*
Quaternary	Recent	Present day monsoon climate, relatively, humid	Present day soils	Trans- Yamuna-Ganga Surface
	Holocene	(a) relatively arid (b) relatively humid (c) Hot and arid climate	airborne sand deposits; buried soil layer, solution and pitting of caliche deposits ; Formation of pedocalic palaeosol, deposition of Liassic sediments;	— —
	Post-glacial	Relatively humid	Trenching and erosion of red-brown-gravel-mottled clay, deep weathering and rubification.	—
	Late glacial, interglacial period	Relatively humid	Trenching and erosion of boulder mottled-clay formation.	—
	Latest Pleistocene glacial period	Cold glacial, arid etc.	Deposition of boulder-mottled clay, frost action	—
	Upper Pleistocene interglacial	Warm tropical humid	Formation of lateritic crusts over Vindhyan bed-rocks	—
				Rewa Erosion Surface
Tertiary	Pliocene	—	High-level laterite	—
	Miocene	—	Tertiary grits and gravels	Panna Surface (rare in Belan basin)
	Oligocene	—	Faulting, middle Tertiary earth movements, faulting	—
	Eocene	—	Faulting, lower Tertiary earth movements	—
Mesozoic (Secondary	Cretaceous	—	—	Kaimur Erosion Surface
Palaeozoic (Primary)	Carboniferous	Cold	Covering of ice sheets	—
	X	X	X	X
			Unconformity	
Algonikan	Purana	—	Upper Vindhyan formation (1) Kaimur series (2) Rewa series (3) Bhander series	—

Erosion Surfaces : The detailed geological structure, tectonic and denudational history of the Belan basin suggest some distinct erosion surfaces but more or less they do not appear to be representatives of complete cycles but they owe their existence to interruptions in the cycles. Some of the morphometric attributes viz. slope, relative reliefs, altimetric frequency histograms and curves and superimposed profiles help in recognizing erosion surfaces. Greater incidence of frequency at 500 ft (150 m), 800 ft (240 m) and 1350-1400 ft (420 m) height groups and parallelism of summit peaks displayed by superimposed profiles indicate the existence of four erosion surfaces. All of these four surfaces have already been identified by R.S. Dubey (1968) m Rewa plateau.

(1) Kaimur surface at the elevation of 1350-1400 ft (420 m) is representative of Bhander surface of Rewa plateau at 600 to 700 m (R.S Dubey, 1968) and 450-500 m surface in the Bijaigarh plateau, Kaimur hills and southern upland of eastern Vindhyan upland (R.L. Singh 1974) and it can be correlated with the 1000-m surface of Chhotanagpur plateau in the east (J.A. Dunn, 1938). This surface belonging probably to Cretaceous period is found over Kaimur hills overlooking Son valley.

(2) Panna surface is not fully represented in the Belan basin except the existence of some patches over Vindhyan scarpland and residuals over Rewa surface. In the extreme south-western part and adjoining eastern projection of Rewa plateau, Panna surface is located at the elevation of 1000 ft (305 m}. The same surface is frequently found over Rewa plateau at the elevation of 400-500 m. Some sort of Tertiary upheavals in the middle Miocene period must have interrupted the Oligocene - Miocene cycle producing a Miocene surface in the form of Panna surface.

(3) Rewa surface at the elevation of 800 ft (240 m) in the Belan basin has its counterpart in the Rewa plateau at 300-350 m. This surface is frequently found in the Belan basin. The post-Pliocene upheaval may be held responsible for the evolution Rewa surface.

(4) Trans-Yamuna-Ganga surface (150 m) may be correlated with the Trans-Yamuna surface of R.S. Dubey, which might have been formed by the destruction of the northern part of the Rewa surface during Pleistocene and recent period. The surface is widely located in the western and central part of the basin.

Ranchi Plateau : The Ranchi plateau, spread over Ranchi district and environs, is a significant physiographic constituent of the Chhotanagpur Higlands and consists of five distinct physiographic units viz.

(i) western highlands or Pat region,

(ii) central Ranchi plateau,

(iii) southern dissected lower plateau,

(iv) eastern lower plateau, and (v) northern escarpments. The present description of denudation chronology and erosion surfaces of the Ranchi plateau is being reproduced from the research paper of Savindra Singh (1978b) entitled, 'Physiographic regions, landforms and erosion surfaces of the Ranchi plateau' published in National Geographer, Vol. 13, No. 1, pp. 43-65.

Ranchi plateau proper (20° 21' N -23° 43'N and 84°E-85°54'E) covering the entire district of Ranchi is the south-western part of the Chhotanagpur plateau. It is separated from Palamau Upland in the west by the N. Koel river, in the north it overlooks the rift valley of the Damodar having wall-like escarpments from which spurs extend upto the river channel, in the east it is bounded half way by the Subaranarekha valley and in the south gently descends to the district boundary where dissected low, forest-clad small plateaus are conspicuous features. The western escarpment (of Patland) registers a steep rise in the west from the surface of the central Ranchi peneplain surface and is characterized by a series of laterite-capped small table-lands separated by deep gorges. The Ranchi plateau has an average elevation of 1066 m in the west, 685 m in the central part and 305 m in the east and south.

The basal rocks of Ranchi plateau belong to Dharwarian ancient sedimentaries which have been intruded by granitic batholiths. These rocks were further metamorphosed by orogenetic movements. Granites and greisses are the main lithological elements of Ranchi plateau and in majority of cases they project above the peneplain surface like mounds and domes. Ranchi plateau has been affected by several phases of earth movements (mainly orogenetic) of different ages and magnitudes. Tertiary period registered successive upliftments i.e. the upliftment of the western Ranchi plateau by 305 m in early Tertiary, second upliftment of 305 m between middle and late Tertiary and third upliftment of 91.44 m.

Denudation Chronology : The geomorphological history of the Ranchi plateau is very much complicated as several chapters of upliftments, intrusion and effusion of lavas, interruptions of fluvial cycles, climatic changes etc. have introduced complexity in the present landform components of the region.

The whole of the Chhotanagpur region experienced glacial phase during upper Carboniferous period but the relics of glaciation are not found over the Ranchi plateau except in the coal fields of North and South Karnapura. This glacial phase must have closed the chapter of earlier cycle and might have obliterated pre-Carboniferous landscapes. The withdrawal of ice sheets might have ushered in a new era of fluvial cycle, which termed as Permian-Triassic cycle might have reduced the region to a peneplain, of which the western part was uplifted upto 305 m, upon which Cretaceous lavas were seated. The Tertiary period registered three phases of uplift and hence interruptions in cycle took place several times. The marginal areas of the plateau characterized by waterfalls, knick points and breaks in slopes in channel gradient and juvenile characters of the rivers where they descend from the escarpments, tell the story of Tertiary upliftments and consequent rejuvenation.

Since Tertiary upliftments, the plateau has been dominated by hot and humid climatic conditions favouring fluvial processes. Even the Quaternary period experienced minor climatic changes in sequential order from hot and arid climate through relatively humid, relatively arid, again relatively humid to present day hot and humid monsoon climate with marked winter dry season. Thus, the present geomorphic features, in majority, are the result of interplay between Tertiary movements causing pronounced upliftment and subsequent fluvial and limited arid processes supplemented with physical and chemical weathering and mass movement of rockwastes.

Erosion Surfaces : The recognition and identification of erosion surfaces is another complicated geomorphological problem. J.A. Dunn (1931) has accepted the western higher plateau as different peneplain. S.P. Chatterjee (1940) recognized four erosion surfaces over Ranchi plateau excepting the higher Netarhat plateau (mesa). R.P. Singh (1969) even accepting three erosion surfaces (e.g. 1. pre-traps erosion surface of the Western Highlands at the height of 915 m, 2. Central Ranchi plateau surface at the height of 610 m, and 3. lower plateau surface at 305 m) amends his proposition as he assigns two reasons for the difference of height between the gneissic surface of the western highlands and central plateau surfaces as he maintained, 'It seems that either the plateau of gneissic rock was higher in the east where it was overlain by lava-flows or the difference in the level marks the extent, of differential erosion since the deposition of lavas-the region covered by lavas having been protected by lava cap, while that to the east was eroded down to the base level' (R.P. Singh, 1969). S.C. Chatterjee is so skeptical about the numbers of erosion surfaces that he has gone to the extent of saying that, 'There is no evidence of the assumption of three or four peneplains. The elevations of the east and the south are due to erosion. The correspondence of summit planes of spurs of both granite - gneiss and softer schists upto the outer edge of the higher central plateau is remarkable feature which shows that both were reduced to the base level' (S.C. Chatterjee, 1945).

No doubt, the altimetric frequency histograms and curves, clinographic curves and superimposed profiles of Ranchi plateau clearly indicate 4 surfaces at the height of (1) 1000-1100 m, (2) 915 m, (3) 610 m, and (4) 305 m, but in no case these are erosion surfaces.

The flat upper surface (1000-1100 m) having accordant summit levels (mesas and buttes) of the Western Higher Plateau (Pat region or Patland) is the result of horizontal disposition of trappean (Deccan) lava (basaltic lava, now lateritic basalt) and the present surface was never reduced to peneplain stage, otherwise the flat-topped mesas might have been denuded into conical hills and ridges as representatives of monadnock stage of landscape development. The junction between gneissic floor and the base of laterites (915 m) has been exposed at several places particularly on the scarp faces and this junction stands 305 m above Central Ranchi Plateau (610 m). This 915 m surface was once a part of the existing 610 m surface of the present Central Ranchi Plateau and the difference of elevation is because of upliftment of the western section of the Ranchi plateau upto 305 m. Following S.C. Chatterjee, it may be forwarded that the eastern and southern lower surfaces at the height of 305 m might have been once a part of main surface of the present Central Ranchi Plateau and later on it might have been reduced to present height because of much erosion due to relative softness of the rocks of the region.

Thus, it may be concluded that the (1) 1000-1100m surface of the Western Highlands is structural in origin ; (2) 915 m surface of the gneissic floor in the Western Highlands is uplifted peneplain; (3) 610 m surface of the Central Ranchi Plateau is true erosion surface, and (4) 305 m surface of the eastern and southern Ranchi plateau is the result of differential erosion. It is obvious that the last three surfaces are, in one way or the other, parts of once an extensive erosion surface formed before the 1st upliftment of the Western Highlands.

Peneplains

Evolution of the Concept : Peneplain represents the landscapes produced at the end of cycle of fluvial erosion. Initially, the term peneplain was used by W.M. Davis to represent featureless plain produced at the end of geographical cycle of erosion. It may be mentioned that W.M. Davis himself admitted that the 'peneplain idea' was not his original one, as is apparent from his own writing, 'At the beginning of this essay I wish to emphasize the fact that the "peneplain idea" was not original with me. The name is my invention, and, as has sometimes happened, the introduction of a definite name for a thing previously talked about only in general terms has promoted its consideration...... It was in Powell's "Exploration of the Colorado River" (1875) that the peneplain idea, along with a number of other important facts and principles, first came to my notice" (W.M. Davis, 1899). It is, thus, evident that the concept of peneplain is based on the works of a few geologists but the term, peneplain, was coined and used first by Davis. A.L. Bloom has further clarified that, 'In an attempt to describe the form of landscapes that have undergone long-continued weathering and erosion in humid climates, Davis (1889) introduced the elegant word peneplain. He used as a root the word "plain" in the geographical context of a regional surface of very low relief near sea level. Realizing that the ultimate base level is the limit of subaerial erosion, which like a mathematical limit may be approached but might not be achieved, he prefixed to the word "plain" the Latin derivation "pene" meaning almost. Thus, peneplain was introduced to the scientific literature as a surface of regional extent, low local relief, and low absolute altitude, produced by long-continued fluvial erosion' (A.L. Bloom, 1978).

It may be pointed out that the concept of peneplain has been the subject matter of severe criticism and controversies right from the stage of its conception to the present day geomorphology dominated by the advocates of non-cyclic development of landscapes.

Characteristic Features : The penultimate stage (old stage) of normal cycle of erosion is characterized by the absence of valley deepening and marked downwasting of reliefs by lateral erosion and weathering processes and flattening of river valleys. Lateral erosion and subsequent backwasting eliminates most of Interstream areas (water divides or interfluves). Valleys become broad and flat characterized by concave slopes of valley sides. Thus, lateral erosion, downwasting and weathering continuously degrade the land resulting into gradual lowering of absolute altitude and water divides. Interstream areas and water divides are remarkably reduced in height and are changed to lowland but they still rise above the surrounding areas. Streams reach their base level of erosion and effective erosion comes to an end and the topography is changed into featureless plain. Rivers adopt meandering courses and are braided. Such featureless plain has been termed as peneplain by W.M. Davis. It may be pointed out that Davis never defined peneplain as a broad and flat surface but he defined peneplain as that surface which was characterized by low reliefs and undulating and rolling surface. It is, thus, obvious that peneplain surface is characterized by upstanding and lowlying reliefs and moderate gentle slope towards the sea so that streams may maintain their flow paths. W. Penck termed the end product of cycle of erosion as endrumpf characterized by concave slope.

The peneplain surface is also exemplified by the presence of some scattered and isolated convexo-concave hills which project above the general surface of peneplains. These relic reliefs have been termed by Davis as monadnocks on the basis of Mount Monadnock of New Hampshire of the USA (North-East Appalachians). The isolated residual hills projecting above endrumpf surface have been termed mosores by W. Penck. The residual hills, when located in groups on peneplain surface, are called unakas (on the basis of Unaka mountain of North Carolina of the USA).

It may be mentioned that a peneplain is a near end product and not end product of cycle of erosion. This fact was also realized by W.M. Davis. He was convinced that it was not possible for a landmass to be eroded down to base level of erosion to become a perfect plain surface. It may be possible theoretically only but cannot be possible in reality.

Time is also an important factor of peneplain formation as it requires long geological period (millions and billions of years) of crustal stability. This is the weakest point of cyclic concept as majority of the geomorphologists are of the opinion that the earth is unstable and hence uneventful long geological period is not possible in reality. The advocates of cyclic concept and peneplain formation do not subscribe to this objection and have presented many examples of perfect peneplains in support of their concept e.g. three peneplain surfaces have been cited as typical examples from the Applachians (USA) viz. Schooley peneplain, Harrisburg peneplain and Sommerville peneplain. Besides, examples of older peneplains have been traced from the Rockies (USA), mountainous areas of Germany, South Africa and central massif of France. It may be pointed out that these peneplains have been so greatly modified by subsequent events that they are not found in their original form, rather their relics are very much identified.

Ranchi plateau (India) presents an ideal example of peneplain. The entire Ranchi plateau was peneplained before Cretaceous lava flow and an extensive peneplain surface was formed at the height of 610 m a.m.s.l. This surface is still preserved on central Ranchi plateau. The western part, popularly known as 'patland' was uplifted by 305 m during Tertiary period and hence the original peneplain surface of 605 m became an uplifted peneplain of 915 m height. It may be mentioned that this surface was buried under lava cover of 154 m thickness during Cretaceous period and hence 915 m surface is still buried under lava cover (now lateritic basalt due to weathering). Thus, the present surface of 1000-1100m of the western highlands (patlands) is not erosional, rather it is structural in character.

Various Peneplains

There are variations in the characteristic features of peneplains in terms of composition of landscapes, processes operating on them, drainage system, debris lying on them, sediments, minor reliefs and hence peneplains also vary spatially. On an average, peneplains are classified into five categories viz. 1. local peneplains, 2. regional peneplains, 3. uplifted peneplains, 4. resurrected peneplains, and 5. partial peneplains.

1. *Local Peneplains*: The process of planation of any landmass by any geomorphic process takes place slowly and gradually through several stages. Consequently, some portion of landmass, under the mechanism of peneplanation, is eroded down to base level while other parts are still under this mechanism. Thus, locally produced featureless plain of subdued reliefs within a larger landmass is called local or incipient peneplain because this represents first phase of peneplanation of the larger geomorphic unit.
2. Regional Peneplain represents the second stage of peneplanation. Regional

peneplains are formed when most of the landmass (geomorphic unit) is eroded down to base level of erosion and several local peneplains coalesce together. Regional peneplain does not represent true peneplain as some portions still remain to be peneplained. It happens that the landmass close to the sea coast is almost eroded down to base level and is thus peneplained but inland areas still have some higher reliefs. Such regional peneplains slope seaward and recede inland and ultimately the entire area is peneplained leaving behind well marked water divides though of subdued relief (low relief). Resistant rocks remain as residuals and project above the general peneplain surface as convexo-concave hills known as monadnocks. Regional peneplains represent those peneplains which are found in situ after their formation, have not been affected by tectonic events (upliftment or subsidence) and have not received any sediment cover (i.e. not buried under debris cover) but there is remote possibility of such condition as the earth is unstable.

3. *Uplifted Peneplain* - The local and regional peneplains, when uplifted due to endogenetic forces, are called uplifted peneplains. Such peneplains are quite old and are markedly modified by subsequent events of warping (during upliftment) and denudation. Most of the older peneplains, identified so far, are examples of uplifted peneplains. Applachian mountains (USA) represent three such uplifted peneplains viz. Schooley peneplain, Harrisburg peneplain and Sommerville peneplain. The Applachians were formed due to Applachian revolution (orogenesis) during Permian period. The dynamic wheels of denudation eroded down the Applachians to base level and thus the highlands were peneplained by the end of Cretaceous (of Mesozoic era) and beginning of Eocene (of Tertiary) periods to form first peneplain, called as Schooley peneplain which was subsequently uplifted in Tertiary epoch and after prolonged denudation second peneplain, known as Harrisburg peneplain, was formed. Harrisburg peneplain (also called as Shenondoah peneplain) was again uplifted and thereafter was peneplained to form Sommerville peneplain. Similarly, Tien Shan mountain of Asia, Scottish uplands, Quebeck plateau (Canada), Rockies of Colorado (USA) etc. are typical examples of uplifted peneplains. The western highlands of Ranchi plateau (Patlands) represent an uplifted peneplain.

Characteristic Features of Uplifted Peneplains — A few terminologies need explanation before describing characteristic features of uplifted peneplains. The area separating two watersheds (river basins) or the area between two streams is called interstream area while the areas having small summits are called summit areas. When there is parallelism between the altitudes of interstream areas and summit areas, this is called accordance and the summits and interstream areas of equal altitudes (accordance) are called accordant summit levels and accordant interstream levels respectively. The uplifted peneplains may be identified on the basis of following characteristic features.

(i) *Accordance in Summit and Interstream Levels* : The regional peneplain is characterized by accordant summit levels and interstream levels. It is, thus, conceived that if a regional peneplain (original) has been uplifted by endogenetic forces, then there must be accordance in the altitudes of summits, interstream areas and crests of interfluves (water divide). It may be possible that the original peneplain might have undergone some modifications and deformations in its original form by warping during the time of its upliftment but the salient features of accordance of altitudes are more or less preserved. The uplifted peneplains must also carry monadnocks of the original regional peneplains. The older uplifted peneplains are characterized by narrow accordant summit levels and interstream levels while recently uplifted peneplains have broader accordant summit levels and water divides.

(ii) *Topographic Unconformity /Discordance* : Topographic unconformity or discordance means younger landforms (i.e. valleys) below and older landforms above. The regional peneplains are characterized by broad, flat and meandering river valleys with very low channel gradient. After upliftment of regional peneplains, rivers are rejuvenated and hence resort to valley deepening (downcutting) with renewed energy (accelerated rate of vertical erosion) and carve out deep and narrow valleys with broad and flat old valleys resulting into the development of valley in valley topography or two storeyed valleys wherein the upper part represents older valleys of previous cycle and the lower part of young valleys (deep and narrow) indicates rejuvenation and initiation of second cycle of erosion. Thus, if a peneplain is an uplifted peneplain, then it must have topographic dicordances as stated above. Secondly, an aplifted peneplain is also characterized by breaks in longitudinal profiles of the rivers. These breaks are called heads of rejuvenation or nick points. The Patlands of Ranchi plateau and Palamau uplands, an example of uplifted peneplain, are characterized by numerous nicks (breaks in slope) in the longitudinal profiles of rejuvenated rivers. For example, there are three nicks in the long profile of the Burha river (a tributary of the North Koel river, which drains into the Ganga river) e.g. Ist nick, in the upper reach, as Burhaghagh waterfalls (142 m), second nick (in the middle course) as Sugabandh falls (12.19 m), and third nick in the lower reach.

(iii) *Presence of Sediment Cover :* The older peneplains are supposed to have thick to thin cover of residual soils and weathered materials (regolith). The materials resting over peneplain surface include rockwastes derived through weathering of water divides and alluvia in the valleys and surrounding areas. The debris cover includes different types of materials viz. laterites, clay, chert etc. The residual soils or rockwastes are seldom found continuously all over the peneplain surface rather these are found in patches.

4. *Resurrected Peneplains* : Some times regional peneplains are buried under sedimentary covers. Such peneplains are called buried peneplains or are submerged under sea water during transgressional phase of the sea and thus receive thick layers of marine sediments. Such submerged peneplains are seen after emergence during regressional phase of the sea. Whenever the surficial (super incumbent) materials are removed through erosion, the buried peneplains are exposed and are called resurrected peneplains. The granite-gneissic peneplain surface (915 m a.m.s.l.) lying below basaltic cover of 154 m thickness (1000-1100 m) is an example of buried peneplain. The 610 m surface of Chhechhari basin of Palamau upland (Bihar) is an example of resurrected peneplain because the Burha river and its tributaries have removed overlying cover through erosion.
5. *Partial Peneplain* : Partial peneplain is in fact incomplete peneplain which is formed due to incomplete or partial cycle of erosion. Incomplete peneplains have been variously termed as incipient peneplains, local peneplains, berms, straths, partial peneplains etc. Following N.M. Fenneman it may be mentioned that if a region undergoes the process of several cycles of erosion, none of the cycles is completed because of a host of reasons viz. earth movements (e.g. upliftment, subsidence etc.), climatic changes, sea-level changes etc. and hence landmass is seldom eroded down to base level of erosion to form true and perfect peneplain as 'the work of next cycle stopped somewhat short of the stage reached in the first, the third fell short of the second, so on.

The Criticism : The postulation of peneplain concept began with initial controversies and still faces stiff opposition from the advocates of non-cyclic theory (i.e. dynamic equilibrium theory) of landscape development.

It may be mentioned that critics of peneplain theory raised a number of objections and arguments against the theory and W.M. Davis (1898) attempted to reply such questions and offered answers to objections point by point. Some of the objections and criticisms of peneplain theory are as follows (cited from Davis, 1898)

1. Certain regions show no trace of peneplanation.
2. The uplands of Southern New England and of Northern New Jersey are not of uniform altitude.
3. The remains of certain peneplains are fragmentary.

4. Certain so-called peneplains are now inclined.
5. The asserted discordance of peneplain surface and rock structure is open to question.
6. The rocks of monadnocks are not proved to be more resistant than those of the adjoining peneplain.
7. No peneplains are now found standing close to base-level.
8. The earth's crust will not stand still long enough for the slow process of denudation to produce a peneplain.
9. No part of the earth reveals even an approximation to a peneplain.

Davis tried to offer statisfactory explanations to most of the objections raised against peneplain theory as follows (A = answer)—

A1. To the argument that, 'one standing upon the crest of one of the mountains of central Maine would hardly find the evenness (of the skyline) sufficient to give the appearance of levelness even to the eye', Davis remarked, 'Its ruggedness is generally so great that it is quite possible that the peneplain explanation does not apply to the greater part of the area (Northern New England). Little wonder that an observer whose attention is given to this mountainous district, under the impression that its mountain tops represent the remnants of a peneplain, should come to discredit such an explanation' (Davis, 1898).

A2. In relation to uneven altitudes of the uplands of Southern New England and Northern New Jersey (Applachians, USA) Davis remarked, 'I should say that the lack of uniformity among the uplands-a fact perfectly familiar to those who accept the peneplain idea is partly the result of tilting, and that for the rest the unevenness of the uplands of today is a natural result of imperfect peneplanation followed by submature dissection' (Davis, 1898).

A3. In relation to fragmentary condition of certain peneplains Davis remarked, 'Its fragemntal condition is most natural; its discovered parts are connected and the lost peneplain is restored by a line of argument that is perfectly reasonable in itself, and that is objected to only because it runs counter to certain views that are held to be established principles in the science of geology' (Davis, 1898).

A4. 'It has been stated that uplift or tilting "is an assumption rendered necessary to explain the difference in elevation of the supposed peneplain, but there is no evidence to prove it unless the peneplain be previously accepted." Davis remarked, 'With this I fully agree. I have repeatedly insisted that it was only by recognizing the existence of a peneplain that uplift or deformation could be determined in certain cases; and that only in this way could certain stages of geological history be discovered...... but it does not seem warranted to conclude that the peneplain theory is invalidated because certain peneplains are now uplifted on a slant' (Davis, 1898).

A5. It has been suggested by the critics that there is 'a very evident general sympathy between the present topography and rock texture' and thus the 'asserted discordance of peneplain surface and rock structure is open to question' as it is not tenable. Davis has observed that, 'the systematic relation of these (uplands in Connecticut, USA) and many other crest lines and uplands suggest a peneplain, and the peneplain thus inferred is strikingly indifferent to the structures that it truncates. It might be argued that the observed discordance of form and structure is of some other origin than peneplanation, but the discordance does not seem open to question' (Davis, 1898).

A6. In response to the objection that, 'the rocks of monadnocks are not proved to be more resistant than those of adjoining peneplain', Davis has remarked, 'I have as a rule given no particular attention to the composition of the monadnock rocks; indeed, it has generally seemed to me reasonable to infer their greater resistance on account of their form' (Davis, 1898).

A7. 'No peneplains are now found standing close to base level.' In response to this objection Davis has submitted that, 'As far as my own understanding of the problem is concerned, it was not at all as an assumption, but as a very surprising corollary, that I came upon the difference between the present and certain parts of the past with respect to peneplanation....... There does not seem to be any severe strain upon the reasonably elastic form of the doctrine of uniformitarianism in meeting the requirements of the peneplain theory' (Davis, 1898).

A8. The earth's crust will not stand still long enough for the slow process of denudation to produce a peneplain'. Davis says, 'Certainly they (lands) do not stand still long enough in certain regions ; witness the manifest effects of uneasiness in the varied and unconformable stratified deposits, or in repeated renewals of dissection in the Alps.... But it has never been my intention to imply an absolute still stand of the earth's crust during an entire cycle of denudation. Any sort of movement that does not cause a distinct dissection of the surface below the peneplain level is admissible.... As to the arguments based on the slow progress of denudation during a brief period of observation or during post-glacial time, I can only reply that a geographical cycle must be so enormously longer than either of these intervals that their evidence is not of value....... Hence the question returns to the verity of the peneplains; whether much or little time is needed to produce them is a secondary matter. Above all, a preconception as to the insufficience of geological time should not in this day be urged as a reason for not believing in the possibility of peneplanation' (Davis, 1898).

A9. 'No part of the earth reveals even an approximation to a peneplain.' In response to this objection Davis remarked, 'It is true that these peneplains (submerged and unconformably buried) are not today standing in the position in which they were formed, that they make a very small part of the earth's actual surface and that they are imperfectly open to observation; but it seems to me that they give strong evidence of the verity of peneplains' (Davis, 1898).

C.H. Crickmay expressed disagreement with the concept of peneplanation of W.M. Davis and proposed the mechanism of panplanation in 1933 to account for the development of landforms in later stage of cycle of erosion. His theory is known as cycle of panplanation. Crickmay first exposed the inherent weaknesses of the processes of peneplanation at least in the late stage of cycle of erosion of W.M. Davis e.g.

(i) the overall profile of residual hills projecting above planation surface (monadnocks) is not convexo-concave as conceived by Davis but is steep and concave in plan;

(ii) shifting of water divides through backwasting and divide wasting through downwearing (downwasting) resulting in progressive decline of slope angle as demonstrated by Davis are not significant processes in the late stage of cycle of erosion;

(iii) most of the planation surfaces are not undulating surfaces rather they are more or less perfect level surfaces i.e. true plains of denudation; and

(iv) dissected peneplains, even in one area, are very often found at various levels or in series, despite the fact that according to Davisian theory the development of a lower surface should inevitably involve the wasting and destruction of a higher surface (R. J. Small, 1970). Crickmay asserted that these discrepancies could be removed if the mechanism of panplanation involving lateral planation (erosion) by meandering rivers was invoked for the explanation of landforms developed in the late stage of cycle of erosion.

According to C.H. Crickmay panplain is the end product of the cycle of panplanation and is formed due to coalescence of flood plains. Crickmay maintained that the process of shifting of interfluves (water divides) and rate of divide wasting had been overestimated and attached undue importance by Davis and his followers. Contrary to this the rate of wasting of interfluves is exceedingly slow which (rate) is further slowed down in the penultimate stage of cycle of erosion because of obliteration of most of reliefs. According to Crickmay the penultimate stage of cycle of panplanation sets in with the cessation of vertical erosion and valley deepening and lateral erosion becomes more active. Lateral erosion causes gradual backwasting of interfluves and hence they are gradually narrowed down and are ultimately destroyed (and obliterated) giving birth to the development of extensive flood plains which are covered with thin layers of sediments.

The flood plains are characterized by steep margins which are cliffed by lateral planation. The margins of flood plains continue to be steepened by lateral erosion upto the end of the cycle. Numerous flood plains ultimately coalesce, when majority of the interfluves are eliminated and thus extensive confluence plain, termed by Crickmay as panplain, is formed. The panplain, thus formed, is characterized by flat and almost even surface which is quite different from the undulating surface of Davisian peneplain.

Crickmay's concept of panplanation could not command wider acceptance by scientific community. In fact, Crickmay overemphasized lateral erosion in the penultimate stage of cycle of erosion. The formation of extensive panplain of level and flat surface due to growth and coalescence of flood plains could not be palatable to many geomorphologists.

8

Slope Development

Meaning and Concept

Slope, defined as angular inclinations of terrain between hill tops (crests) and valley bottoms, resulting from the combinations of many causative factors like geological structure, climate, vegetation cover, drainage texture and frequency, dissection index, relative reliefs (and of course denudational processes, including weathering, mass wasting and mass movements of rockwastes, erosion and transportation of eroded materials downslope) etc. are significant geomorphic attributes in the study of landforms of a (fluvially originated) drainage basin (Savindra Singh and R. Srivastava, 1975). Slope is, thus, upward or downward inclination of surface between hills and valleys and form most significant aspect of landscape assemblages. Except plains and terraces slopes are always perceptible and are more significant in mountainous regions. In fact, morphological characteristics of a given region are determined by slopes of that region because physical landscapes are the result of combinations of slopes. Not only this, slopes introduce variations and complexity in the landscapes of a region. This is why the study of different aspects of slopes has always been a focal theme in geomorphology.

Two terms, related to slopes i.e., angle of inclination and inclined surface, need clarification. Angle of inclination of the surface of a slope means slope angle or simply slope whether hillslope or valley side slope whereas inclined surface means slope profile of either hillslope or valley side slope. The characteristic angle or characteristic slope 'is that which is most frequently identified under particular conditions of rock type and climate' (R.J. Chorley et. al, 1985) and has maximum frequencies of slope angles of hillslope.

The study of slope in geomorphology is given more importance because, 'not only slopes do comprise the greater part of the landscape, but as an integral part of the drainage system they provide water and sediment to streams. Therefore, hillslopes are an important component of the complex landscape that forms a drainage basin' (R.J. Chorley et. al 1985). The slope study involves classification, development and evolution of hillslopes. Besides, slope profile study based on field measurement of slope angles and instrumentation of processes acting on hillslope has become more important for quantitative analysis of hillslopes and construction of theories and models of their development and evolution. Historically, the researches on slope study and analysis may be grouped in two phases. Initial phase was dominated by interpretation of different aspects of hillslope and valley side slope development on the basis of field observation. Davisian (slope decline) and Penckian (parallel retreat and slope replacement) models of slope development belong to this phase of qualitative study of slopes. Modern phase is dominated by quantitative analysis of slopes based on data derived from topographical maps and aerial photographs, measurement of slope angles in the field, and instrumentation of processes (weathering, mass wasting and movement, and erosion) acting on the hillslopes. The works of R.A. Savigear, A. Young, Waters etc. belong to this phase. Besides, significant contributions in slope study have been made by J.L. Rich (1916), C. K. Wentworth (1930), A.C. Lawson' E. Raize and J. Henry (1937), R.E. Horton (1945), H. Baulig (1935), W.C. Calef (1950), W.C. Calef and R. New Comb (1953), A. Wood (1942), L.C. King (1953), A.N. Strahler (1950), C.A. Cotton, O.M. Miller and C.H. Summarsoo (1960), S.A. Schumm (1956,1967), S.A. Schumm and M.P. Mosley (1973), M.J. Selby (1982), R.J. Small and M.J. Clark (1982), Young (1961, 1963, 1972), Savindra Singh and R. Srivastava (1975, 1977), Savindra Singh (1979), Savindra Singh and R.S. Pandey (1982, 1987), Savindra Singh and S.P. Agnihotri (1982) etc.

The Classification

There is marked variation in opinions regarding the classification of hillslopes because of confusions regarding the elements (segments) and form of slopes. Convex, free face, rectilinear and concave segments are in fact elements of hillslope profile and the forms of slope but a few geomorphologists have recognized them as types of hillslopes. It may be pointed out that no hillslope may be only either rectilinear (straight), or convex or concave rather it comprises more than one element (segment), though one element may be most extensive and dominant. Generally, some geomorphologists classify slopes for simplification into cliff slope (free face or scarp slope), rectilinear or straight slope, convex slope and concave slope, and if more than one forms are present, they call them composite slope. But this classification is not justifiable because these are not slope types (slope forms) rather these are slope elements (slope segments). So let us first examine the characteristics of slope elements.

Elements of Slope

If the longitudinal profile of a hillslope or coastal slope is considered, it becomes apparent that there is no uniformity in the slope profile in relation to slope angles from hill tops to valley floors or from cliff crest to sea shore. The entire slope profile is punctuated by the, presence of convexity, concavity, rectilinearity and free face. These distinctive segments of slope profile are called slope elements or slope segments. It is not always necessary that all the slope profiles will comprise all the four elements but an ideal hillslope profile is that which consists of summital convexity, free-face, rectilinearity and basal concavity. The existence of free face (cliff) element depends on the presence of resistant strata (in the case of hillslope) or undercutting (cliffing) at the slope base (in the case of coastal slope).

(1) *Summital Convexity* : The convex segment is found at the hill crest (hill top) and this element is called summital convexity (or simply convex element). Convex slope, if convex element dominates the hillslope profile, is also called as waxing slope because it grows in height upward and in dimension downward but this terminology of W. Penck is not justifiable because not all convex slopes necessarily grow in height and dimension (size). Most of the summital convexity develops because of denudational processes e.g. rainwash or soil creep, this is why convex slope is also called upper wash slope.

(2) *Free Face* : Free face element of hillslope represents wall-like precipitous slope of bare rocks and is devoid of any debris. The slope is so steep and precipitous that no weathered materials can rest on it. The element or segment is called slope of derivation because there is instantaneous downslope transport of materials. The free face element is subjected to backwasting which facilitates parallel retreat of this element.

(3) *Rectilinear Element* : The straight or linear segment of hillslope profile between upper free face and lower concave element is called rectilinear element. It is also termed as constant or uniform or regular slope because of uniform nature of slope angle (i.e. gradual increase upslope or gradual decrease in slope angle downslope). Because of presence of rock debris this element is called debris slope. The angle of rectilinear segment is controlled and determined by angle of repose of debris and it is also called as repose slope. It may be pointed out that this terminology is confusing and gives the impression that rectilinear element is always the result of accumulation of debris but this is not always true because some times rectilinear segment is originated due to denudation and exposure of bare rock surface. It may be mentioned that there may be a thin layer of debris but this is not stationary rather it is mobile i.e. it is in constant motion down slope. This is why rectilinear element is called debris controlled slope.

(4) *Concave Element* : The basal segment of an ideal hillslope profile is always characterized by concave element. The slope angle decreases as the segment of basal concavity increases (in length). This is why this segment is called waning slope. This element is also called valley floor basement slope or lower wash slope. Basal concavity is generally originated because of active denudation mainly by rainwash, rill and gully erosion. This segment may be covered with debris or may be of bare rocks.

Slope Elements based Classification

It may be pointed out that theoretically it is conceived that an ideal hillslope profile consists of all the four elements viz. summital convexity, free face, rectilinear and basal concavity but in practice it is not always possible. The escarpments of Bhander plateau (M.P.) present a typical example of ideal hillslope profile consisting of all the four slope elements but summital convexity is not well pronounced. Slightly upslope dipping beds of massive Vindhyan sandstones of 152 m thickness resting as caprock on underlying shales and mudstones have given birth to steep and precipitous scarps as free face element. The thick deposits of shales and mudstones below sandstones have been responsible for the development of rectilinear and basal concave elements.

It may be mentioned that since development of ideal hillslope profile with four elements depends on local conditions (e.g. nature and disposition of rocks, structure of rocks, nature of denudational processes, climatic conditions, vegetation etc.) and hence some times one or two elements do not develop. For example, free face does not develop in the regions of low or subdued reliefs. On the other hand, if resistant and weak rocks are alternated in a hillslope, there is repetition of free face (on resistant rocks) and rectilinear elements (on weak rocks). The sequence of these four elements changes with the passage of time during cycle of erosion. For example, summital convexity is dominant element in youthful stage while rectilinearity becomes more pronounced in mature stage of cyclic development of landscape. Waning slope or concave element becomes most extensive during old stage. Thus, we may label a slope on the basis of a single element such as convex slope, rectilinear (straight slope), free face slope or concave slope but in reality this is rarely possible. Any slope profile consists of at least two elements and instead of mono-element slope there is composite or compound slope which may be of different types depending on varying combinations of slope elements.

(i) Convexo-concave slope

(ii) Convexo-rectilinear-concave slope

(iii) Free face-rectilinear-concave slope

(iv) Convex-rectilinear-free face-rectilinear concave slope

(v) Convex-rectilinear-free face slope (coastal area, basal free face is developed due to basal cliffing by sea waves), and so on

Genetic Classification

Genetic classification of hillslopes involves the mode of their origin and development. Thus, on the basis of mode of genesis hillslopes may be classified into three types.

(1) Tectonic slope is generally formed due to tensional and compressional forces resulting into the formation of fault scarps through faulting in crustal rocks, and due to tilting of rock beds. Scarp slope is ideal example of tectonic slope. It may be mentioned that original scarp slopes are modified by denudational processes and hence they become compound slopes.

(2) Erosional slopes are generally formed due to erosion by rivers, glaciers and sea waves. Erosional slopes developed by rivers undergo se quential transformation and changes in different stages of cycle of erosion i.e., convex slope in juvenile (youth) stage, rectilinear slope in mature stage and concave slope in old stage. Most of the erosional slopes by the streams are valley side slopes. Sea waves form steep cliff slope through basal undercutting of coastal rocks. Free face and pediment slopes are formed by fluvial erosion in arid and semi-arid regions.

(3) *Slope of Accumulation* : The slopes formed due to deposition of eroded sediments by different denudational processes are called aggradational slopes or slopes of accumulation. For example, slopes of alluvial fans and cones formed by rivers, sand dunes by wind, morainic ridges deposited by glaciers etc., are aggradational slopes. Volcanic cones also come under this category.

On the basis of stage of formation slopes are classified into (i) primary slopes and (ii) secondary slopes. Primary slopes are formed due to erosion by streams, glaciers, sea waves etc. V shaped valleys with convex plan of gorges and canyons, cliff slopes carved out by sea waves,, steep slopes of U shaped glacial valleys etc., are examples of primary or erosional slopes. Tectonic slopes (e.g. fault scarp slope) also fall in this category. Secondary slopes represent minor slopes developed at the base of primary slopes. These slopes are formed due to surface erosion and weathering. Slopes of talus cones are formed due to accumulation of debris/ scree coming down from the hillslope as a result of mass movement of rock wastes.

Quantitative Classification

Quantitative classification involves classification of slopes into certain types on the basis of slope angles which are derived either from topographical maps and aerial photographs or on the basis of measurement of slope angles with suitable instruments in the field. A. Young classified slopes on the basis of slope angles of slope profiles into seven major categories as follows.

1. Level to gentle slope
   (a) level slope (0°—0.5°)
   (b) almost level slope (0.5°—1.0°)
   (c) very gentle slope (1.0°—2.0°)
2. Gentle slope (2.0°—5.0°)
3. Moderate slope (5.0°—10.0°)
4. Moderately steep slope (10.0°—18.0°)
5. Steep slope (18.0°—30.0°)
6. Very steep slope (30.0°—45.0°)
7. Precipitous to vertical slope (45.0°—90.0°)
   (a) precipitous slope (45.0°—70.0°)
   (b) wall-like slope (70.0°—90.0°)

Approaches of Slope Development

In the initial stage the study of slope development was based on qualitative information received from field observations but the problems of slope continued to increase. The major problems of slope development before the geomorphologists include the following aspects,

1. varying forms of slopes,
2. most slopes are compound slopes as they consist of more than one slope elements,
3. slope is a three dimensional feature,
4. denudation balance of slope,
5. progressive changes in the form and gradient of slopes with time,
6. relationships between the processes of mass movements (slides, creep, flow etc.) and slopes,
7. parallel retreat of slope and grading of slope profile,
8. influence of geological structure and lithology (rock types) on slope forms,
9. relationships between climates and slope forms etc. The problems of slope evolution and development may be approached in three ways vie. theoretical approach, experimental approach (laboratory experiments) and empirical approach (i.e. field

studies). These three approaches may be fused in two categories vie. (1) slope-evolution approach and (2) process-form approach.

(1) Slope-evolution approach involves the study of historical development of hillslopes. Davisian model of slope decline comes under this approach. There are certain problems which are faced by the geomorphologists while attempting to describe historical evolution of slopes e.g. 1. It becomes necessary to find out original and initial form of present day hillslope but this becomes very difficult because of lack of requisite evidences and hence the study is based on speculations, deductions and thoughtful reasoning. Most of the geomorphologists generally assume that the initial form of slope might have been vertical cliff or scarp, which might have undergone subsequent changes and transformation due to weathering, mass movement and erosion and resultant progressive decline in slope gradient to attain the present form. 2. The dating of slope development poses a serious problem before the investigators. In order to solve this problem attempts are made to study the mechanisms and processes of changes in slope form and slope decline. Slopes are placed in time sequence on the basis of study of slope profiles in the field and their relative dating. This method may be applicable to only those areas where slope decline has not been obstructed. This method would not be applicable to valley side slopes of such rivers which have formed incised meanders due to rejuvenation because the topographic discordance (the segments of valley sides being older than the lower segments) disturbs the sequential pattern of slope evolution in time sequence. In spite of these problems W.M. Davis postulated his concept of slope decline (slope evolution) wherein slope undergoes sequential but progressive changes (decline in slope gradient) from very steep slope during youth through rectilinear (uniform or straight slope) slope of moderate angle during maturity to concave slope of very low angle during old age. On the other hand, some geomorphologists have opined that it is not always necessary that all the slopes undergo progressive decline in slope gradient through time rather some slopes maintain their angles through parallel retreat. According to advocates of dynamic equilibrium theory time plays no significant role in slope development because this process is time-independent and slope processes play most important role in their development.

(2) Process-form approach is based on the concept that there is direct relationship between slope types and slope forms, and geomorphic processes active on slope profiles or between slope types and denudation. The intensity of denudational processes is influenced by structure, rock types, climatic conditions, vegetation cover, relief etc. The relative variations in the influences of these controlling variables introduce variations in slope forms and slope types. For example, limestones and clay in humid climate will differently respond to fluvial processes and hence resultant slope would greatly vary e.g., soil creep would be most active on limestones resulting in the formation of convex form

of slope whereas clay rocks would be subjected to intense rainwash resulting in the development of concave slope form.

This approach also suffers from some difficulties and problems. 1. Slope forming processes operate so sluggishly that it becomes very difficult to observe, measure and record their rates of action 2. It is also difficult to isolate such processes which affect slope development because not all processes acting on hillslope profile are concerned with slope development. For example, W. Penck recognized soil creep, rainwash etc., as transporting agents which transport the weathered materials downslope. In fact, these processes erode and transport only weathered regoliths and not the solid rock surface. 3. Most of the present day slopes do not appear to be the result of present day geomorphic processes, rather they represent relict features of formerly developed slopes. The advocates of process-form approach argue that the relationship between present day processes and slopes is without doubt established at least in the areas of soft rocks which are subjected to continuous denudation. 4. If the concept of association between denudational processes and slope form is accepted then it should also be accepted that there must be close relationship between climatic conditions and slope forms (i.e. slope forms vary from one climatic type to other) because climate determines the nature of denudational processes. For example, fine materials (debris) are produced due to more active chemical weathering in humid climate because of abundance of rainwater. On the other hand, coarse materials (debris) are produced due to more intense mechanical weathering in arid climate. But the critics of the concept of climatic geomorphology have refuted this claim and maintain that it is not necessary that distinctive slopes are produced in different climates. For example, climatic geomorphologists take pediments of concave surface as representatives of arid and semi-arid climates but such pediments are now also found in sub-humid tropical as well as humid temperate climates. In fact, except periglacial climate different combinations of all the four elements (convex, free face, rectilinear and concave) are found in almost all the climates. This is why L.C. King propounded the concept of climatic uniformitarianism.

It may be concluded that slope forms are controlled not by a unique factor rather these are controlled by several factors such as geological structure, rock types, vegetation cover, different types of weathering, dip angle, mass wasting and mass movement of rockwastes (creep, slides, flow etc.), earth movements, reliefs etc. It may be pointed out that in certain environmental conditions one of these factors may be most dominant but that cannot be the only factor of slope development.

It is now generally believed that geomorphic processes (of weathering, mass movement, erosion etc.), without doubt, control the forms of slopes and their development but question arises as to whether slope form is a function of mono-process or poly-processes.

Thus, there are two approaches to the study of process-form approach of slope studies viz. (i) mono-process concepts, and (ii) poly-process concept.

It may be mentioned that significant slope forming processes are soil creep (solifluction, earth creep, debris creep), slides (earth slides, debris slides etc.), flows (earth flows, debris flows, mud flows); slope-erosion by overland flow, rain splash, rainwash, rill and gully erosion etc.

(i) Mono-process concept is based on the basic tenet that a particular denudational process or process of mass movement produces a distinctive slope form. For example, convexity and concavity are considered to be the outcome of soil creep and rainwash respectively. According to N.M. Fennman (1908) convexo-concave hillslopes are the result of differential actions of surface runoff and resultant overland flow i.e., the crestal areas of hillslope are least eroded by surface runoff because of limited amount of water (due to limited space) and erosion tools and hence rainwash only smoothens the hill crest so as to give it convex plan but the intensity of erosional work of runoff increases downslope bc cause of substantial increase in surface runoff (coming from upslope) and debris with the result the basal segment of hillslope is subjected to maximum erosion which ultimately forms concave slope. Thus, the convexo-concave hill slope profile is the result of least erosion at the hill-crest and maximum erosion at the basal segment by surface runoff. This concept of Fenneman suffers from the weakness that it does not accommodate the process of soil creep which is most dominant in humid regions because a sizeable portion of rainwater infiltrates in the porous and permeable rocks and thus motivates soil creep on one hand, and infiltration of rainwater lessens the intensity of erosional work of surface runoff because of reduced (due to more infiltration) runoff and resultant overland flow on the other hand.

G. K. Gilbert (1909), on the other hand, attached more importance to soil creep in the development of convex slope of such a hillslope profile which has a cover of soils. The water- soaked soils move downslope. As we move away from hill-top downslope the amount of soil to be moved downward increases. Thus, there is minimum soil cover at the hill crest while it is maximum at the lower segment of the slope because of soil coming from upslope. According to G.K. Gilbert the total removal of soils from the hillslope may be possible only when the slope becomes steeper downslope so that the slope load (soils) may be totally removed through the process of soil creep. In such condition the hillslope assumes convex form so as to allow effective movement and evacuation of debris from the slope. It may be pointed out that in such situation (increased transport of soil layer downslope) there would be accumulation of huge amount of debris at the slope base and hence the basal slope will become concave rather than convex. Gilbert's assumption, that soil layer on slope profiles is always mobile, is not always true because it is not always necessary that soil layer on hillslope, even in humid regions, is mobile.

Lawson's View

According to A.C. Lawson (1932) summital convexity of hillslope, like N.M. Fenneman, is formed by rainwash erosion but Lawson's concept differs significantly from Fenneman's concept as the former postulated most effective rainwash at and near hilltops due to under-loaded surface runoff while the latter (Fenneman) postulated least effective rainwash at the hill-top because of less volume and velocity of surface runoff. According to Lawson erosion by surface runoff at and near hill-tops becomes most active because surface runoff carries less debris and hence there is more available energy of surface runoff for erosion which very effectively rounds the crestal part of the hill giving it convex slope at the top while the lower segment of the hillslope becomes concave due to deposition of sediments coming down from upslope segments. As per postulation of Lawson there is two distinct zones in hillslope profile viz. 1. zone of erosion, and 2. zone of aggradation. The upper segment is subjected to maximum erosion by rainwash wherein lune shaped materials are removed from the crestal part and are transported downslope and are deposited at the foot-hill of the slope.

The continuous removal of material through rainwash results in the development of summital convexity representing the zone of erosion. The radius of the curvature of longitudinal profile of hillslope increases because of continuous deposition of materials at the slope base coming down from upslope. This results in the formation of basal concavity. This lower zone is in fact erosionless zone because of the fact that, according to Lawson, erosional energy of surface runoff decreases downslope (in spite of downslope increase in the volume of runoff) because the amount of debris increases downslope and hence surface runoff becomes overloaded and sluggish. R.J. Small has remarked, 'A corollary of Lawson's theory is, therefore, that with time relief is diminished, divides are wasted, and slope angles steadily decline, much as envisaged by Davis' (R.J. Small, 1970). Several questions are raised against the concept of Lawson. The most effective erosion by surface runoff at the hill-tops inspite of low volume of water' due to limited space, and least erosion or no erosion at the slope base inspite of increased volume of surface runoff, with high flow velocity, as claimed by Lawson, cannot be easily digested. The development of basal concavity by deposition of debris, as claimed by Lawson, is not of common occurrence, instead, active erosion (mainly by rill and gully erosion) at the slope base is responsible for the development of basal concavity because the erosive energy of surface runoff increases downslope due to substantial increase in the volume and velocity of runoff. It may be opined that concavities may be depositional or erosional depending on local conditions.

(ii) Poly-process concept of the study of slope development is nearer to reality because there is every likelihood that more than one denudational processes act on slope profiles. H. Baulig postulated the poly-process concept of slope development in 1950 and refuted the concept of 'one process-one slope form' (mono-process concept). According to Baulig different denudational processes operate independently on slope profile (they do not operate together). It may be possible that a particular process may be most active on one segment of the hillslope while other process may be more active on the other segment of slope. Soil creep (solifluction) and rainwash have been recognized by Baulig as most active slope forming processes. Soil creep becomes more active on upper segment of convexo-concave hillslope in humid temperate regions with the result summital convexity is developed (due to soil creep). On the other hand, more rills and gullies are developed due to large volume of water on the lower segment of hillslope and hence basal segment is subjected to intense fluvial erosion by rills and gullies while soil creep is significantly retarded, with the result basal concavity is formed (due to fluvial erosion, simply called as rainwash). The relative significance of soil creep (active on upper segment) and rainwash (active on lower segment) changes with time. There is marked lowering of relief (relief reduction) due to continued denudation and hence the process of soil creep becomes sluggish (with time) but rainwash still continues. This results in extension of length of concave unit upslope at the cost of summital convexity and ultimately concave form spreads over entire hillslope. It appears from Baulig's concept that slope processes operate in opposite directions but this is not the truth because most of denudational processes operate together on hillslope.

Contrary to the above concept some geomorphologists have opined that soil creep and rainwash are not slope forming processes rather they are agents of transportation of debris produced by weathering of slope materials.

Slope Evolution Models

The origin (evolution) and development of hillslopes and valley-sides slopes may be approached in three ways viz. (i) theoretically (theoretical models), (ii) experimentally (experimental design models) and (iii) empirically. Any model or theory of slope development must have solutions of a few important questions related to slope development e.g. (i) whether slopes undergo parallel retreat and maintain their slope angles or (ii) there is progressive decrease in slope angles with time, (iii) whether hillslopes are subjected to downwasting or backwasting etc. Based on these two basic issues of slope development (i.e. parallel retreat and constant slope angle - dynamic equilibrium, and progressive slope decline with time) there are two distinct schools of thoughts which are based on contrasting views of W. Penck and W.M. Davis. Significant models (theories) of slope development include slope decline model of W.M. Davis, slope replacement model of W. Penck, slope evolution model of A. Wood, hillslope cycle model of L.C. King, concept of A.N. Strahler, Process-response model of Fisher and Lehmann, Process-response model of A. Young etc. Following A. Young (1972) all the theories of slope development may be grouped in 3 broad categories viz. (1) Slope decline theory (2) Slope reslope development by R.A. Savigear, concept of placement theory, and (3) parallel retreat theory.

'Slope decline — the steepest part of the slope progressively decreases in angle, accompanied by the development of a convexity and concavity'.

'Slope replacement — the maximum angle decreases through replacement from below by gentler slopes, causing the greater part of the profile to become occupied by the concavity. The concavity may be either smoothly curved or segmented'.

'Parallel retreat — the maximum angle remains constant, the absolute lengths of all parts of the slope except the concavity remain constant, and the concavity increases in-length; the angle at any (vertically constant) point on the concavity decreases. This type of evolution may be sub-divided into parallel retreat with a free face and without a free face'. (A. Young, 1972)

Theory of Slope Decline

Davis' theory of slope decline has its roots in his essays on 'the convex profile of badland divides' (1892), 'the grading of mountain slopes' (1898), 'the geographical cycle' (1899), 'piedmont bench lands and primarumpfe' (1932) etc. Like the cyclic development of landscapes Davis' hillslope and valley side slope also undergo the process of cyclic development wherein there is progressive decline in slope angle and sequential change in slope form from youth (convex form) through mature (rectilinear or uniform slope form) to old (concave form) stages. Steep convex slope evolves during youth stage of cycle of erosion because of active downcutting resulting into valley deepening and weathering processes. There is very limited slope retreat and practically there is no decline in slope angle rather it is increased. Lateral erosion dominates over vertical erosion and divide summits are eroded (downwasting of water divides) during mature stage. Thus, downwasting of water divides results in decrease in slope angle (and hence slope decline) and slope profile of smooth curve is formed. Slope becomes graded because at each point on the slope profile the gradient is such that weathered debris may be transported (removed) downslope. Because of continued lateral erosion and downwasting of water divides (and hence marked lowering of relief) there is marked slope decline in old stage so that the general slope form becomes concave and nowhere slope angle becomes more than 5 degree.

Davisian model of slope evolution includes three aspects viz. rounded convexity of hill tops and interfluves, graded waste sheets on slope profiles and graded valley sides, which Davis tried to explain in terms of their significance and origin. According to Davis summital rounded convexity results from the action of soil creep in humid climate. 'He described qualitatively the mechanism of movement in creep, by alternate dialation and contraction of the soil under the influence of gravity, and correctly interpreted outcrop curvature' (A. Young, 1972). Soil creep is motivated because of rainwash the intensity of which increases downslope. 'Reasoning on priori grounds that surface wash increases in volume downslope, he supposed that, near drainage divides, the ratio of creep to wash is large. Creep (soil creep) produces 'rounded contours', and is responsible for convex profile of divides' (A. Young, 1972).

The weathered materials existing on slope profiles (both hillslope and valley side slopes) has been termed as waste sheet which constantly moves downslope by the agents of transportation under the influence of gravity force. The nature, rate and magnitude of downslope transfer of debris depends on the nature and amount of debris and transporting capacity of denudational processes. When the transporting capacity of the transporting agents (actual available energy of transporting agents) equals the required energy (force) to transport the materials i.e., when the available energy for transportation of debris and the work to be done (debris to be transported down slope) are equally balanced, the layer of waste sheet of debris on slope profile is called 'graded waste sheet'. In youthful stage upper slopes are so steep that available energy (transporting capacity) far exceeds the work to be done (debris to be moved downslope) and hence debris is quickly removed and transported downslope because of the fact that materials supplied through weathering and erosion of upslope segment are far less than the transporting capacity of the denudational processes due to steep gradient of slope. It may be pointed out that condition of graded waste sheet begins from the base of the slope and gradually proceeds upslope. The entire slope profile is graded (i.e. transporting capacity equals the total amount of debris to be transported downslope throughout the profile length) by the attainment of old stage of cycle of erosion and the whole slope profile from hill top to the base is covered with sluggishly moving rock waste. Davis has defined graded waste sheet as follows, 'a graded waste sheet...... is one in which the ability of the transporting forces to do work (to transport) is equal to the work that they have to do ...... from any point on such surface a graded slope leads the waste down to the streams. At any point the agencies of removal are just able to cope with the waste that is there weathered plus that which comes down from further uphill' (W.M. Davis, 1899). Thus, there is balance between supply of debris and removal of debris at all points of graded slope profile.

He has further observed that in the initial stage graded slope profile has steeper gradient and thin veneer of coarse debris (waste) but as the stage of cycle of erosion advances the gradient of graded slope profile declines, debris becomes finer and debris thickness increased. Similarly, the gradient of valley side slope also decreases as the stage of cycle of erosion advances with time. The valley side slope is dominated by summital convexity (in the upper part of the valley sides) and basal concavity. With the advancement of cyclic time (stages) the radius of the curvature of valley side slope profile increases because of gradual increase in the length of convex and concave segments. This again denotes flattening of slope profile and resultant decline in slope angle.

Like cycle of erosion Davis' model of slope evolution also depends on crustal stability for longer duration after the initiation of cycle. A. Young has remarked that, 'the Davisian model has, however, great flexibility combined with internal logical consistency, which renders it unassailable by critical attacks with its own weapons, those of abstract qualitative reasoning... Davis' theory is a clear statement of a type of slope evolution, and is capable of being translated into quantitative terms. Consonant with its relation to the cycle of erosion, the theory implies that slope form is time-dependent; possibly this is the feature that offers an opportunity of indirect testing by field observation' (A. Young, 1972)

Theory of Slope Replacement

It may be mentioned, at the outset, that Walther Penck is most misunderstood geomorphologist of the world because of his (i) incomplete work due to his untimely death, (ii) his obscure composition in difficult German language, (iii) ill defined terminology, (iv) misleading review by W.M. Davis, (v) faulty translation in English language, (vi) some contradictory ideas etc. Previously Penck's concept of slope development was widely understood as 'parallel retreat of slope' but his real concept on slope evolution came to light, as slope replacement model, after English translation of Penck's work i.e. 'Die Morphologische Analyse' by Czech and Boswell and again by M. Simons in 1962.

The main goal of Penck's morphological system was to find out the mode of development and causes of crustal movement on the basis of exogenetic processes and morphological characteristics. In other words, Penck sought the interpretation of diastrophic history of regions on the basis of the interpretation of present-day landform characteristics. The reference system of Penck's model is that the characteristics of landforms of a given region are related to the tectonic activity of that region.

Penck's 'morphological system' has two important parts viz. (i) manner (mode) of action of processes, and (ii) deductive model of slope evolution. According to Penck the mode of operation of exogenetic processes depends on the characteristics of regoliths on hillslope, mechanism of weathering and surface processes. He attached more importance to the process of mass movements in comparison to other surface processes. He has analysed the mechanism of processes with sound reasoning and has attempted to establish relationship between mechanism of processes and characteristics of forms. According to Penck the mechanism of surface processes is affected and controlled by the properties of forms and in turn the latter also undergo changes and transformation. Penck has described seven properties of forms and 3 properties of processes.

Various Properties

1. *Degree of reduction* of regolith means degree of breaking of regoliths into fine materials. It may be mentioned that fine textured soil has higher degree of reduction than coarse textured stony soils. The degree of reduction (comminution) depends on the rate of weathering and denudation. It may be mentioned that Penck used denudation as the process of removal (transportation) of slope debris and not as erosional process.
2. *Mobility of regolith is* that property of regolith which determines downslope movement of slope materials by denudational processes (of mass movement). Higher the mobility of regolith, more the downslope transfer of slope materials by denudational processes and vice versa. Mobility depends on rock properties and degree of reduction.
3. *Thickness of regolith* means thickness of mobile or static cover of loose materials on hilislope and it is directly related to degree of reduction (the higher the degree of reduction of regolith the thicker the regolith and vise versa) and is inversely related to rates of denudational processes (the higher the rate of denudational process i.e., the higher the rate of removal of slope materials downslope, the thinner the cover of regolith and vice versa).
4. *Exposure of slope surface* means degree to which the rock; under regolith are uncovered and are made available: for reduction by weathering. Exposure of rocks under regolith is inversely related to the thickness of regolith.
5. *Rock property* includes all those aspects of lithology and geological structure which influence the degree of reduction and mobility of regolith.
6. *Climate :* Only those aspects of climate are considered which influence the rate of reduction and denudation (removal of materials).
7. *Slope angle* becomes independent variable in the initial stage of slope evolution but later on it becomes partly dependent on the rate of denudation.

Processes and Forms

1. *Reduction* means mode of breaking of regolith into fine particles. The mechanism of comminution of regolith is controlled by weathering processes while the rate

of breaking of regolith depends on rock characteristics, climate and exposure of slope surface of solid rocks.

2. *Denudation is* related to removal (transport) of regolith materials downslope. The rate of denudation of materials (removal of materials) is dependent on the rate of weathering, mobility of regolith, climate and slope angle.

3. *Renewal* of exposure means uncovering of solid rock surface, under regolith, and its exposure to the processes of reduction (weathering processes). It depends on the rate of denudation or rate of removal of materials from slope surface (direct relationship). It may be clarified that renewal of exposure is not a process in itself but is the result of removal of regoliths because bare rock surface, after the regolith materials are removed and are transported downslope, is exposed to weathering processes.

Though W. Penck has described interrelationships between the properties of forms and processes but A. Young has further attempted to rationalize these interrelationshps after making certain modifications and has transformed Penck's model of slope evolution into process-response model of slope evolution.

Basic Premises

(1) The form of hillslope is dependent on the relative rates of vertical erosion by streams at the slope base and denudation (removal or transportation of slope debris downslope or from the slope base).

(2) The form of hillslope is not directly controlled by slope processes because these act as agents of removal (transportation) of weathered slope debris downslope. The major role of denudational processes (transportational processes) is to expose bare rock slope surface (by removing slope debris) for weathering processes.

(3) The retreat of slope unit backward depends on the gradient of hillslope. Steeper gradient facilitates more active retreat than gentle gradient.

(4) The slope retreat occurs in parallel manner which results in the replacement of lower segment of slope profile by new unit of gentle gradient.

(5) 'Flattening of slopes always takes place from below upward'.

(6) The form of valley side slopes of streams depends on 3 rates of stream erosion i.e. 1. accelerating rate of erosion, 2. decelerating rate of erosion, and 3. constant rate of erosion, which produce convex, concave and rectilinear slope forms respectively.

(7) There is uniform rate of weathering of rocks of slopes.

(8) As the required mobility is attained, the removal of weathered materials begins and the rate of removal (denudation) matches with the rate of weathering.

(9) There are angular breaks of slope (breaks of gradient) having base-levels and the development of gradient of slope segment is independent of lower basal levels of breaks of gradient. A. Young (1972) has remarked, 'Penck does not say, as some commentators have assumed he does, that a smoothly accelerating rate of erosion will produce an angular break of slope.'

A. Young (1972) after making certain modifications in the relationships, interrelationships and interactions in the properties of forms and processes, as described by Penck, has presented Penckian model in the form of process-response model'. Firstly, he differentiated between dependent and independent variables. Secondly, he identified three closely related pairs of properties e.g. (i) degree of reduction and mobility, (ii) thickness of regolith and exposure, and (iii) denudational processes and renewal of exposure. The effects on the rate of processes become operational for the whole year or for lesser duration. Such effects are called short-term effects. The factors affecting regoliths are of medium-term effects (1,000 to 10,000 years) while the factors affecting slopes are of long-term effects (10,000 to 100,000 years). Young has further suggested that for the testing of Penckian model on quantitative base rock properties should be divided into two independent variables of rock property viz. (i) those which influence degree of reduction and (ii) those which determine mobility of regolith. Similarly, climatic variables should be grouped into two independent variables viz. (i) those which influence degree of reduction and (ii) those which influence rate of denudation (removal of weathered materials).

Slope Evolution Model of Wood

A. Wood presented his concept of slope development in 1942 through his research paper entitled, 'the development of hillside slopes' published in the Proceedings of Geological Association of London. Wood's concept of slope development includes (a) parallel retreat of slope and (b) adjustment between rate of weathering and rate of transportation of weathered materials (downslope removal of materials). He selected a cliff slope with free face element for the purpose of elucidation of his concept. The cliff slope might have been formed either tectonically (faulting) or erosionally. The free face is subjected to weathering which causes its retreat through backwasting. The weathered debris is transported downslope and accumulates at the slope foot (base). The accumulation of debris extends upslope with the result the lower segment of free face slope is continuously covered with debris and thus is buried under debris cover. Consequently, the length of free face decreases (i.e. height of free face decreases). Thus, the lower segment of free face buried under debris cover becomes constant slope. A convex rock slope is formed under debris covered constant slope. This simple condition may get disturbed when a stream becomes active on debris constant slope. Due to stream action weathered materials are removed from the lower segment of slope. According to Wood's assumption the streams are competent enough to remove all the materials coming from upslope segment of free face due to weathering. In other words, there is balance between the available energy (transporting capacity of streams) and the work to be done (weathered debris to be removed). This situation (of equilibrium) leads to the development of constant debris slope i.e., the length of debris slope remains constant. The regular retreat of free face through backwasting resultants in regular extension of constant slope but no debris can rest on this extended constant slope, which is called denudational constant slope. Thus, constant slope consists of two segments e.g. (i) lower aggradational segment and (ii) denudational segment which is also called as slope of derivation or transportation slope and is an example of rectilinear slope. As the mechanism of retreat of free face through backwasting caused by weathering continues, the rectilinear segment extends upslope and ultimately reaches the divide summit when the free face element completely disappears or say when the process of retreat of free face is finally completed.

The disappearance of free face results in the rounding of divide (interfluve) summit and the development of summital convexity. With the march of time lower segment of constant slope becomes concave. Thus, the entire slope profile consists of three elements i.e. summital convexity, rectilinearity and basal concavity and an ideal convexo-rectilinear-concave slope is formed. The evolution of slope still continues and the summital convex and basal concave units extend at the expense of rectilinear unit. Ultimately, the rectilinear unit disappears and convexo-concave slope is formed. With the march of time this slope is subjected to gradual wasting and decline in slope angle.

Evaluation : The model of slope evolution of A. Wood is, in fact, geometrical in appearance and indeed is very interesting but it is unable to solve some problems and confusions (regarding the slope development) in reality. It is not likely that convexity and concavity would appear in the last phase of slope development. Wood fails to present convincing argument for this situation. The transformation of constant debris slope into concave form may be accepted, at least conceptually, but it is very difficult to digest the arbitrary assumption of transformation of convex rock slope buried under constant debris slope into concave slope form. If the constant debris slope is once transformed into concave form due to stream erosion, then this concave form is superimposed on the convex rock slope which is eroded down due to removal of overlying debris.

Theory of Hillslope Cycle

The King's concept of hillslope evolution is, in fact, related to his broader scheme of landscape evolution, which includes the concepts of 'river cycle' (1951), 'hillslope cycle' (1953) and 'landscape cycle' (1962). It maybe mentioned that King's model is based on the concepts of A. Wood (1942), T.J. Fair (1948, 1949), W. Penck (1924) and R.E. Morton

(1945). In fact, King did not survey slopes in the field and did not measure slope making and controlling processes but based his model on field observations alone. 'King's opinions lack a basis of detailed field observations; no measurement of either process or form appears in his published work.

Although this is severe indictment of any study of the natural environment, it could be maintained that he was concerned to present a broad view, based upon reconnaissance observations on a world scale previously unparalleled in geomorphology, and that this is no bar to subsequent detailed investigations of particular aspects' (A. Young, 1972).

After a careful observation of South African landscape he observed that humid temperate climate and areas could not be 'normal' for landscape development, as claimed by W.M. Davis, because most of the topographic features had been modified by Pleistocene glaciation; and most of the present day landforms are relict features developed in the past periglacial climate. Thus, according to King, subtropical semi-humid regions are 'most normal' for landscape evolution. He also rejected the basic tenet of climatic geomorphology that 'the variations in slope forms and slope elements depend on climatic types i.e., slope forms vary from one climatic type to the other' and maintained that, 'our thesis will be that the basic physical controls of landscape remain the same in all climatic environments short of frigid or extremely arid' (King, 1957).

According to King a standard (ideal) hill slope consists of all the four elements e.g., summital convexity, free face, rectilinearity and basal concavity. Such ideal hillslope is the natural products of normal processes of slope evolution involving fluvial process (flowing water) or mass movement or both. He further maintains that full development of slope elements depends on local conditions i.e., resistant and strong bedrocks and bold and sufficient reliefs. If either of these conditions are absent, free face (scarp) and debris slope do not develop and thus a waning convexo-concave slope is formed.

The evolution of hillslope is, in fact, related to the concept of pediplanation cycle of King. According to King pediplain formation is the result of twin processes i.e., scarp retreat and pedimentation. In the initial stage of hillslope development, the scarp face experiences parallel retreat due to backwasting caused by weathering of exposed rocks. This parallel retreat of scarp controls the evolution of entire hillslope. According to King the debris slope just below the scarp or free face does not extend upslope and hence it is neither capable of obscuring nor destroying the free face element. It means that there is a balance between the supply of debris from upslope and removal of debris downslope on this section i.e., debris slope. Similarly, there exists balance between debris supply and debris removal at hill crest and hence it remains constant. The gradual parallel retreat of free face and rectilinear elements (debris slope) results in the formation of pediment of concave plan at the base of the hillslope. As the parallel retreat of free face and debris slope continues, their lengths decrease and pediments extend upslope at the cost of rectilinear and free face elements. Ultimately, upper slope elements mainly free face disappear and pediments extend upto hill crest. Thus, an extensive erosion surface of concave slope is formed. This surface is called pediplain, which is, in fact, the product of coalescence of several pediments. Thus, the whole process of development of pediplain and the entire process of slope evolution during this period is called 'pediplanation cycle' and 'hillslope evolution cycle'.

King has further maintained that if the scarp is absent due to lack of resistant rocks and bold relief then the process of slope evolution would be entirely different from the process of parallel retreat and pediplanation. There will be regular decrease in maximum slope angle and extensive convexo-concave slope is produced. According to King the hillslope with scarp is considered normal because scarp (free face) is the active element of slope retreat. Wherever scarp is absent, the present slope form is not the result of present geomorphic processes. Based on this assumption King has maintained that extensive convexo-concave slopes of the temperate regions of the northern hemisphere are, in fact, relict forms and are not the result of present day processes rather they are the result of past periglacial processes. These have not been formed due to scarp retreat rather these have developed because of weathering and erosion of weak rocks.

Evaluation : King's model of slope evolution is, in fact, not original concept of his own rather it is amalgamation of views of earlier workers like A. Wood, T.J. Fair, R.E. Horton etc., and is a story-like deductive and impracticable but interesting description. His concept is based on extensive reconnaissance observations alone. No field survey and instrumentation of slope making processes have been undertaken. King has used assertions rather than evidences in support of his model. 'Evidence from the work of others is selectively mustered in support of preconceived views. An example is the argument put forward involving the hydrological distinction between laminar and turbulent flow of water; it is asserted that turbulent flow on the upper slope element gives place to laminar flow on pediment. This has never been experimentally verified. Where there exists clear field evidence contradicting a view, the circumstances are dismissed as not 'normal' (A. Young, 1972)'. A. Young (1972) has further remarked, 'his writing lacks objectivity. The value of King's work is in setting up hypotheses for testing; they are not proven'.

If King's views are considered in Indian context, some of his ideas become relevant and are validated. The author after having close observations and selective measurement of slope angles in certain parts of foreland of Indian peninsula viz. Chhotanagpur highlands, Rohtas plateau (Bihar), Kaimur ranges, Rewa plateau, Bhander plateau (M.P.) etc. has noticed precipitous scarps (free face element) which are undergoing the process of parallel retreat due to backwasting caused by physicochemical weathering of well jointed massive sandstones (of scarps) resting over shales and siltstones forming debris or rectilinear slope. Bhander plateau having imposing escarpments on its eastern, south-eastern and southern sides consists of all the four slope elements (crestal convex, free face, rectilinear and basal concave elements) wherein scarp or free face element is subjected to parallel retreat. A few of the detached hills having sandstone capping are the examples of mesas characterized by flat top but steep scarps on all sides. These scarps are experiencing parallel retreat due to backwasting.

The Savigear Concept : Unlike other geomorphologists British geomorphologist R.A. Savigear (1952) postulated his concept of slope development on the basis of field evidences rather than on deductive reasoning. His concept, thus, is nearer to reality and ground truth. He closely observed and studied the characteristics of a few slope profiles at the head of the Carmarthen Bay in South Wales and postulated that both parallel retreat and slope decline played important role in slope development. Some times these processes work together in a single region of uniform environmental conditions. He identified one such slope profile in the eastern part of the Carmarthen Bay which consisted of basal free face (scarp slope), middle rectilinear element with slope angle of 32° and upper limited summital convex element. The angle of rectilinear slope is such that the weathered and eroded materials are removed from the rectilinear slope due to gravity and are transported down the free face or scarp slope. The materials coming at the base of free face are instantaneously removed by sea waves and thus basal free face slope is maintained but it experiences parallel retreat. On the other hand, there are convex-rectilinear-concave slopes in the western part of the Carmarthen Bay where basal concave slope has developed because of accumulation of debris at the slope base as the materials coming from upslope cannot be removed by sea waves because the sea has receded and thus sea waves do not reach the slope base.

According to Savigear rectilinear slope at constant angle of 32° can be maintained only when there is effective removal of materials from its base but if there is no removal of materials from slope base, the debris coming from upslope accumulates at the base of rectilinear slope and thus the lower segment of slope is protected from weathering and erosional processes whereas the upper part of the slope profile continues to retreat due to weathering. This mechanism effects decline in slope angle. He has also inferred that the development of summital convexity takes place at the later stage of slope development. 'It does seem possible, to judge from the evidence in Carmarthen Bay, that both parallel retreat and slope decline are tenable theories, and that two mechanisms can exist together in one small area in response to different conditions of slope-foot erosion or deposition' (R.J. Small, 1970).

It may be mentioned that difference in lithological conditions may also lead to parallel retreat as well as slope decline in close proximity. For example, the Bhander scarps and detached hills (mesas) having sandstone capping resting over shales (Satna district, M.P.) are experiencing parallel retreat while the Sharda Pole hill located very close to Bhander scarps is experiencing slope decline as the sandstone capping has been finally removed due to parallel retreat and thus the hill is subjected to downwasting which has produced convexo-concave slope.

Strahler's Concept : The concept of A.N. Strahler, an American geomorphologist, on slope development is based on field data obtained through measurement of slope angle in carefully selected areas of uniform conditions in terms of geological structure, lithological characteristics, climatic conditions, vegetation, soil, relief etc. He first measured maximum slope angles in a region of uniform environmental conditions and then calculated mean maximum slope angle after averaging maximum angles. On the basis of close perusal of maximum slope angles and mean maximum slope angles he postulated that there was very little deviation of individual maximum slope angles from mean maximum slope angles in a given locality. According to Strahler the hillslopes have developed uniform maximum slope angle so that slope materials may be effectively transported downslope. Such slope, from which debris is removed downslope easily, has been termed by Strahler as equilibrium slope which is controlled not by a single factor but by all those factors which are involved in the development of slopes. Slight change in any one of the factors may disturb the equilibrium slope and may warrant readjustment of equilibrium condition. Strahler drew three inferences from his empirical study of slope development :

(i) in any one locality there is uniform maximum slope angle on all retreating slopes.

(ii) it is not necessary that all such slopes are of the same age in terms of evolution, and

(iii) it is more likely that slopes having uniform maximum slope angles undergo the process of parallel retreat rather than slope decline.

After a detailed study of slope profiles in fluvially originated drainage basins Strahler attempted to establish correlation between valley side or ground slope and channel gradient or slope. According to him valley side slopes are positively correlated with channel slope i.e. wherever valley side slope is steep, the channel slope is also steep so that weathered and eroded materials coming from the valley side slopes may be effectively transported by the channel downstream. Conversely, gentle valley side slopes generate relatively lesser debris and hence the channel gradient also becomes gentle. Following Strahler such condition (positive relationship between valley side slope and channel slope) may be possible only when the slope making and controlling factors remain constant. Change in any of the factors may disturb such condition and relationship. For example, two valley side slopes have uniform angle but one is covered with vegetation while the other one is devoid of vegetation. The vegetation-covered valley side would generate less debris because of low degree of denudation and hence channel gradient would be gentle in comparison to vegetation-free valley sides and associated channel gradient because in this case valley sides are subjected to more denudation, generate more debris and hence channel gradient would be relatively steeper so that all, the loads may be transported downstream.

According to Strahler, if the channel gradient is steepened either due to rejuvenation or due to change in discharge-load ratio consequent upon climatic change, the steepening of valley side slope becomes more common than parallel retreat. He further maintains that wherever the landscape evolution is controlled by base-level, wherein slope gradient decreases with time, then the possibility of slope decline cannot be ruled out. The valley side slope becomes steeper only in Caose conditions where river channel touches the slope base so that all the debris coming from upslope may be transported by the channel downstream easily. On the other hand, if the channel is away from the base of valley sides, the materials coming from upslope are deposited at the base and hence the valley side slope becomes gentle.

It may be mentioned that the views of Strahler very much fall in line with those of R.A. Savigear in that parallel retreat and maintenance of steep slope angle is possible where materials coming from upslope are removed from the slope base and slope decline becomes operative where materials accumulate at the slope base.

Fisher-Lehmann Model : O. Fisher (1866) and O. Lehmann (1933, 1934) postulated process response model of slope evolution. It may be pointed out that process-response model is related to the mechanism of slope forming processes and their response i.e., changes in slope form during a given time span. First, Fisher applied this model for the explanation of cliff recession in 1866. The initial form of the cliff slope is wall-like steep slope with precipitous scarp (free face element) and there is accumulation of debris at the foot of the cliff. A.C. Lawson also deduced the process of cliff recession in 1915 while O. Lehmann added a few more variables for the explanation of development of cliff slope in 1933 and 1934 e.g. angle of cliff, scree angle, ratio of rock-scree volume. They deduced the concept of parallel retreat of scarp (free face) section of the cliff slope. The model of cliff recession is generally known as Fisher Lehmann model (A. Young, 1972).

According to this model it is assumed that the initial form of slope is rectilinear-free face wherein initial height is denoted by h and slope angle by R. It is further assumed that the surface at the top of cliff and at its bottom is flat and plain. The entire section of free face (scarp) is exposed to weathering processes. Because of impact of weathering processes a weathered layer of uniform thickness is detached and falls down in a unit time. This dislodged weathered layer of scree accumulates at the foot of the slope. Thus, the slope angle of newly formed rectilinear slope due to accumulation of scree is less than the angle of original (initial) free face (R). Cliff undergoes the process of continuous parallel retreat due to backwasting caused by weathering of exposed rocks of free face element. A rock core is formed due to cliff recession and accumulation of scree at the foot of cliff. This rock core is buried under scree cover. It is assumed that there is no change in the form of rock core. The height of free face (h) gradually decreases because of gradual accumulation of scree.

This model envisages that if the cliff face is subjected to gradual parallel retreat and there is accumulation of scree at its (cliff) base then the form of rock core buried under scree would be convex. The lowest point of convex rock core becomes tangent to initial cliff angle (R). When the process of parallel retreat is complete and the entire free face (scarp) is removed due to parallel retreat and scree accumulation, the highest point of convex rock core (H) becomes tangent to scree angle (x). The changes in initial cliff angle (P) have insignificant control on the shape of scree-covered rock core as it remains convex in plan. If the cliff angle (3) is less than 90°, the lower segment of rock core is relatively of gentle slope. On the other hand, change in scree angle (x) has greater impact on the form of rock core. As the angle of scree slope increases, the steepness of rock core also increases and vice-versa.

If the entire mass of scree is removed during parallel retreat of cliff, the resultant relict rock surface produced by parallel retreat gives birth to rectilinear slope which is called denudational slope. It may be mentioned that the steepness of rock core depends on the steepness of initial cliff angles i.e., if the initial cliff slope angle is very steep, the convexity of rock core would also be very steep.

Process-response Model

A. Young (1963) and F. Ahnert (1966 and 1970) independently developed process-response model of slope development but the models appear more or less identical. Young has constructed his models 'on consideration of the interaction of processes and form, calculated by the method of integration'. The following description of Young's model is, in fact, literal reproduction of Young's version from his book 'Slope' (A. Young, 1972).

Basic Assumptions

Model 1. (i) Unimpeded basal removal (of scree), but no basal erosion (of accumulated scree by flowing water).

(ii) Slope retreat subject to control by removal (of accumulated scree).

(iii) Removal of regolith (scree) entirely by surface transport.

(iv) Rate of surface transport (is) proportional to sin ϕ and distance from crest (of cliff).

Model 2. (i) Unimpeded removal (of regolith), but no basal erosion (of accumulated regolith by flowing water).

(ii) Slope retreat subject to control by removal (of accumulated regolith).

(iii) Removal of regolith entirely by surface transport.

(iv) Rate of surface transport (is) proportional to sin ϕ and distance from crest.

Model 3a. (i) Unimpeded basal removal (of regolith), but no basal erosion (of accumulated regolith by flowing water).

(ii) Slope retreat subject to control by removal (of accumulated regolith).

(iii) Removal of regolith entirely by direct removal.

(iv) Rate of direct removal (is) proportional to sin ϕ or (Model 3b).

Model 3b. (i) Unimpeded basal removal (of regolith), but no basal erosion (of accumulated regolith by flowing water).

(ii) Slope retreat subject to control by weathering.

(iii) Rate of weathering (is) proportional to sin ϕ (rate of surface transport immaterial).

Model 4. (i) Unimpeded basal removal (of regolith), but no basal erosion (of accumulated regolith by flowing water).

(ii) Slope retreat subject to control by removal (of accumulated regolith).

(iii) Removal of regolith partly by surface transport and partly by direct removal.

(iv) Rate of surface transport and rate of direct removal both proportional to sin ϕ.

Model 5. (i) Unimpeded basal removal (of regolith), but no basal erosion (of accumulated regolith by flowing water).

(ii) Slope retreat subject to control by removal (of accumulated regolith).

(iii) Removal of regolith entirely by direct removal.

(iv) Rate of surface transport (is) proportional to sin ϕ and to regolith thickness.

(v) Rate of weathering — conversion of rock to regolith (is) proportional to sin ϕ and inversely proportional to regolith thickness.

(vi) Regolith thickness (is) proportional to total regolith weathered upslope of a given point, and inversely proportional to rate of downslope transport at the point.

Model 6. (i) Impeded basal removal.

(ii) Where ground loss occurs, slope (is) subject to control by removal; where ground gain occurs, slope (is) subject to control by accumulation (of regolith).

(iii) Removal of regolith entirely by surface transport.

(iv) Rate of surface transport (is) proportional to sin ϕ and distance from crest (of cliff).

A. Young (1972) has claimed that, 'although constructed with dimensionless parameters, actual slope dimensions may be applied to slope models, together with observed or assumed rates of action of processes. The time required for a given amount of profile evolution may then be calculated.' Young has attempted to calculate the time required for the development of slope profile of two different heights on the basis of assumed rates of downslope transport and direct removal of debris. In the case of model 1, if the height of initial slope is 10 m and assumed downslope transport of regolith is 0.5 cm'/cm/year and there is no direct removal of regolith from slope base, then the rounding of angular junction and development of convexity would occur in 15,00,000 years but if the height of initial slope is 100 m then the time required for the rounding of angular junction and development of convexity would be 150,000,000 years. If the rate of downslope transport of regolith is increased to 3.0 cm/cm/year, then the rounding of 10 m - high slope would be complete in 2,60,000 years and the time required for the rounding of 100 m - high slope would be only 26,000,000 years. In the case of model 5, if the assumed rate of direct removal of regolith from slope base is 0.005 cm per year for 10 m- high slope, the time required for the development of slope would be 1,00,000 years and 1,000,000 years for 100 m high slope. If the assumed rate of direct removal of regolith is increased to 0.02 cm/year, then the development of 10 m-high and 100 m-high slope would require 25,000 years and 2,50,000 years respectively.

Displacement of Slope Materials

Slope failure refers to displacement of slope materials and their downslope movement in different forms caused by a host of factors e.g. inclination and intensity of jointing, rock strength, rock complexes, climatic factors mainly through water pressure within the slope, groundwater conditions, infiltration of rainwater and melt-water, weathering factor, human factors (deforestation, construction of roads and buildings, dams and reservoirs etc., mining, blasting etc.).

Types of slope failure include falls (rock falls, slab failures, toppling failures, rock avalanches, soil fall, earth fall, debris fall), slides/landslides (block slide-block glide, rock slides, debris slides, soil slip, slumping), flows (earth flow, debris flow, mud flow), creep (soil creep, rock creep, talus creep), spreads (rock spreads, debris spreads, earth spreads) etc.

Erosion and Hillslope Processes

Hillslope processes include those erosional processes which modify slope form such as creep, overland flow, rainsplash, rainwash, through flow, rill and gully erosion etc. Besides, slope processes also include those transportational processes which remove debris downslope. Soil creep or solifluction is, in fact, transportational process which facilitates mass but slow movement of moistened and loosened soil soaked with water which acts as lubricator. Soil creep plays major role in rounding the hillslope mainly crestal rounding and in giving convex slope form to hill crest (i.e. summital convexity). Overland flow or popularly known as surface runoff is considered to be effective agent and process of slope erosion and slope modification. The surface flow of water resulting from rainfall parallel to the slope is called overland flow or surface runoff. The rainwater received through rainstorms on hillslope moves in different directions viz. (i) a portion of rainwater is retained on slope surface and is held there until it is finally evaporated, (ii) a portion infiltrates downward and moves vertically to meet groundwater table, (iii) remaining water moves laterally parallel to hillslope to become overland flow or surface runoff. It may be mentioned that surface runoff occurs when the rainwater and rainfall intensity exceeds the infiltration rate which is dependent on the nature of surficial material, rainfall amount, intensity and duration. Initially, infiltration rate of rainwater is very high and no surface runoff is generated if the soil is dry but if the duration of rainstorms becomes relatively longer and rainfall intensity is also high then infiltration rate decreases and surface runoff is generated. The erosion of hillslope and its modification by surface runoff depends on its velocity which depends on the nature of slope surface in terms of surficial materials and vegetation cover, surface roughness, volume and depth of water and slope angle. According to Manning equation the velocity of overland flow or surface runoff may be expressed as follows

$$V = \frac{1.49}{n} d^{2/3} S^{1/2}$$

where v = flow velocity

d = depth of flow of water

n = roughness index

S = slope inclination

Significant Mechanisms

The erosive power of surface runoff depends on flow velocity of overland flow, nature of surficial materials, slope angle, vegetal cover etc.

Rainsplash refers to erosion of loose superficial matters by the direct impact of falling raindrops. It may be mentioned that the intensity of splash erosion depends on the kinetic energy of falling raindrops. If the slope is covered with vegetation, the splash erosion becomes minimum to nil because the falling raindrops are intercepted by vegetation cover and hence the kinetic energy of drops is markedly reduced. Thus, it is apparent that rainsplash becomes more effective on vegetation free slopes where falling raindrops directly hit the surface with full kinetic energy.

'The combination of water on slope (surface runoff) plus raindrop impact produces an intermediate value of erosion. In other words, the sheet of water dissipates the raindrop energy to some extent but raindrop impact, nevertheless, provides sediment to the flowing water and so erosion continues' (Chorley et. al, 1985).

Through flow, also known as interflow, refers to lateral flow of infiltrated water down the slope below slope surface. It becomes more effective on those slopes which are densely vegetated because dense vegetation allows maximum infiltration of rainwater as falling raindrops are intercepted by forest canopy and thus rainwater reaches the ground surface very slowly in the form of aerial streamlets which carry rainwater through leaves, twigs, branches and stems of trees to the ground surface. This infiltrated rainwater after reaching a comparatively impermeable layer moves down th‿ slope below the upper surface as through flow or interflow which causes slope failure through slumping and collapsing of upper slope surface. Besides, interflow also carries matters in suspension and solution. Sometimes, interflow leads to the process of piping which causes the formation of tunnels underneath and after some time the roof collapses during strong rainstorms.

Rilling, also known as rill and gully erosion, is another significant mechanism of slope erosion. Rilling refers to linear erosion of superficial loose materials as well as rock beds through concentrated linear flow of water known as rill. Rilling is more effective on vegetation free steep slopes. The pattern of rills depends on the steepness of slopes. They become dendritic on gentle slopes but become parallel on steep slopes. The intensity of rill erosion increases downslope. Most of basal segments of hillslopes right from the Himalayas to the Eastern and Western Ghats, where vegetation has been partly or completely removed, have been greatly rilled and gullied and the process is becoming more and more vigorous with the exposure of underlying rocks and divagation.

Following Chorley et. al it may be concluded that 'hillslope erosion can take many forms, the simplest two being erosion by creep and overland flow, but there can be combinations of the two processes and even seasonal changes in the intensity of each process' (Chorley et. al, 1985).

9

The Departure

Scientific Geography

In the preceding survey of the development' of geographic thought in Germany, we gave but brief attention to radical suggestions calling for major changes in the concept of geography that had little if any effect on subsequent thought. It is true however, as Leighly has recently commented, that methodological-controversies that are in themselves of minor importance may suggest significant conclusions. The fundamental attitude that underlay the proposals of Frobel in the 1830's and of Gerland a half century later is one to which expression is frequently given in discussions-particularly oral discussions on the nature of geography. Since Frobel and Gerland both expressed their views fully in print, and the subsequent works of both give us some basis for judging the significance of their methodology to their practice, it may repay us to examine their programs more closely, in the light both of the background which produced them and the results to which they led.

Frobel's Programme : We have previously noted that Frobel's programme was based on a very limited knowledge of the field which he sought to reform. Prior to his coming to Berlin in 1832, his preparation for geography consisted of several years experience in topographic mapping and map drawing, university studies in natural sciences other than geography, a general reading of travel books, and finally-work which chance threw in his way and which need impelled him to accept-the preparation of a geographic handbook *on* Peru, Bolivia and the Plata countries, needed to complete the twenty volume *"Handbuch der Erdbeschreibung,"* edited by Gaspari, Guts Muths, and others. In many respects this series ("one of the traditional encyclopedic treatises," as Leighly correctly calls it, in another connection was of a type that antedated Gatterer as well as Ritter, and, though

Frobel attempted to introduce short treatments of the natural regions of each country in addition to the standardized division by minor political units, one does not wonder that he mentions it many years later with little pride. By the end of the year or two needed to complete this job, he knew what was wrong with regional geography, so that his program for reform in geography appeared in the same years as his first publications in the field.

Frobel was therefore perhaps the first of the many students of the past hundred years who have attempted to construct a science of geography based, not on a study of geography itself, but rather on a comparison with the "natural sciences." Starting with the two naive assumptions, namely that "geography" or "*Erdkunder*" must correspond to a literal translation of its name, and' that it "can only be a natural science," Frobel thought his way through in print-not without some confusion—to what may be regarded as a logical program. This program divides geography into two parts. Geography as a natural science consists of separate studies of each of the features of the earth, classified by kind. Following, in part, Hunboldt's point of view, as well as his terminology, he included the study of man as a feature of the' earth in his "natural science," in terms of all objective facts of man and the earth that represent relations between them.

On the other hand, those aspects of geography that could not be fitted into his frame of a "natural science" including what we have called "aesthetic geography," the relation of historical phenomena to geography, and, apparently, regional geography-he recognized as a form of "applied geography" or "historical-philosophical geography." The two points of view are to be developed by their own methods, the former following the analytic the latter the synthetic method. Each can be built into a separate system, but the two apparently cannot be combined. Geography is, therefore, not merely "dualistic" but rather, one would suppose, is completely separated into two kinds.

To what consequences did this view of geography lead in Frobel's work? Plewe presumes that there should have been no difficulty in pursuing a program so clearly laid out *i.e.*, in the "scientific" half of geography—but Frobel's later work was not known to him. Indeed in all the studies of the history of geography that I have seen, Frobel's name disappears completely after its one sporadic entry. Although it recurs in periodical reviews; in later decades, one hardly recognizes it as belonging to the same man; only: his autobiography provides the connection.

During the decade following 1832, Frobel taught geography and mineralogy in Zurich and apparently devoted a considerable part of his time to the preparation of a full study of systematic geography that was to lead to a theory of earth science, but this was never to be completed. Overlooking a study in crystallography a part of his work in his second field of instruction—the only distinctly geographic research publication of his life, other than his initial *Handbook of Peru, etc.*, was a study of certain remote valleys in the Walliser Alps, in which he was chiefly concerned to map the topography and to study the ethnography of the population-a remnant of early Celtic tribes, as his linguistic comparisons led him to conclude. This was hardly, one may suppose, a study in the strict "natural science" sense of geography.

After a decade of teaching at Zurich, Frobel resigned from his position to devote himself to revolutionary propaganda in the political field and never returned to academic life. After serving in the abortive Frankfurt parliament of 1848, and narrowly escaping the firing squad in Vienna, he went in exile to America, where he spent ten years in various activities, chiefly as a roving journalist. The two-volume story of his travels-several times across the continent, as well as in various Central American countries, but not in the South American countries of which he had written in his first geographic work-was favourably reviewed by Neumann in the periodical of the geographical society of Berlin and provides interesting reading today.

Although a large part of the work is concerned with social and political problems in the United States, and the author specifically indicates that it is not to be considered as a scientific work, he nevertheless has provided a large amount of geographic material. One short chapter is a systematic study of the mountain systems of western North America—the only study that Frobel ever made that could incontestably be included in what he had defined as scientific geography. (Whether this was a more significant contribution to the physical geography of North America than Neumann thought, I am not in a position to judge; it is interesting to note that it was first published in the *California Chronicle*. More valuable are the many fine descriptions of areas, which could fit into his division of "applied geography"; these offer the historical geographer descriptions of our South-west in the middle of the past century made by a trained geographer. The informed reader will note the student of Humboldt in the description of the scenery of the Limpias Valley as "a wonderful harmony and unity of physiognomic elements out of which the landscape is formed".

Publication of Frobel's Work : Returning to Germany, Frobel published in the following years a large number of political studies, some of which are clearly based on a geographic foundation. Notable is his short study of the relation of America to the European political-geographic situation—a study which at the end of his life he regarded as more important than all else that he had written, and which Humboldt found valuable. Frobel is therefore characterized by Sporer—in the same study to which Leighly has called our attention, because of its criticism of Ritter and his followers-as a "skillful historical-geographical chess player" . . . "the virtuoso in the art of combining history and geography". Indeed one may say that the same Frobel, who in his first geographic phase appears as a herald of the late nineteenth century "purely scientific" geography, in his later "most important" study 'Stands as the forerunner of the twentieth century, post-War, school of *Geopolitik.*

The remainder of Frobel's long and active life-ten years or more as publicist and special adviser to various German governments in Vienna, Stuttgart, and Munich, and some sixteen years in the consular service of the German Reich, in Smyrna and Algiers—does not appear to have produced any contributions to geography.

One cannot consider the career of this extraordinary man without wondering what position he might have attained in the field of geography had he remained in it. In view of his undoubted capacities and energy it seems probable that he might well have influenced its developments as much as any other student of his time, but that his predominant interest in human phenomena would have permitted him to confine himself to that part which he classified as "scientific" appears most doubtful. Whatever one may conjecture on that point, the fact is that his revolutionary program for geography, based neither on his own work nor on a study of the history of the field, had no more influence on his own subsequent publications than on the development of the field in general.

The radical program that Gerland proclaimed for geography bore even less relation than Frobel's to the author's previous work; and likewise had but little influence on his subsequent studies; before and after, his major field of work was in ethnographical studies which to some degree could be considered as ethnographic geography. In this case, also, the biographical background is significant; it may be found in the sympathetic study by Sapper. The fact that Gerland has had no university training in geography was not unusual. Most of the men who developed the field in the latter part of the century had come into geography after being trained in mathematics, geology, zoology, or history. In most cases, however, the shift resulted from the interest that they found in the field and was represented first by publications in the field of geography. Gerland's relation to geography, however, had apparently been limited to the surveys of ethnographic work that he wrote for the *Geographisches Jahrbuch* and the teaching of courses in geography that fell to his lot in various secondary schools, when, in 1875, the new German university of Strassburg called "the forty-two year old ethnologist and philologist" to the chair of geography. Although Sapper regarded the experiment as successful, we have contrary evidence in Gerland's own statement. In explaining his proposals for radically changing the nature, of the field to which he had so belatedly been called, he wrote that the unscientific nature of geography as commonly understood had been "embittering his life and career". If one reads the whole of Gerland's lengthy statement, one cannot but question whether he had been able to understand the field he was called upon to teach-whether the problem, in other words, was not-simply the personal problem involved in the contrast between his particular way of thinking in science and that of the field of geography as Humboldt and Ritter had developed it.

Whatever may be the explanation of Gerland's personal problem, his programmatic essay of 1887 did not solve it. Throughout the next twenty years of teaching he continued to include the geography of man in his courses. Among his writings of this period but little can be found that would be included in the field to which he had proposed to limit geography. As a major work, Sapper notes his *"Geographische Schilderung des Reichslandes Elsass-Lothringen,"* published in 1894, a work that included consideration of the human population, but "on principle" excludes political geography'—Hamlet without the Prince of Denmark one may say, since the only justification for considering this area as a unit is to be found in its political geography.

One can hardly read Gerland's examination of Kant's work in geography and anthropology-a lecture course given in 1901 and published a few years later-without wondering whether he had not actually given up his earlier thesis, even in theory. It seems but a weak echo to hear him urge "all geographers to take to heart the word of the great philosopher," namely that "physical geography, a general abstract of nature [including, as we have noted, man] is the foundation of history and all other possible geographies". Ratzel might have said as much, and Richthofen in fact did. At about the same time, Sapper informs us, Gerland planned to write four great works: "geophysics as the geography of inorganic earth, the geography of plants. The geography "of animals, and sociology as the geography of man".

The reader may well ask why we should further consider a program for radical limitation of the scope of geography which had no basis in the author's previous work and the influence on his subsequent work. Nevertheless, Gerland's program is significant as the most thoroughgoing attempt to outline a science of geography in terms of the concept of science developed by a particular group of sciences, from "purely logical" considerations, regardless of the historical development of the field. In this respect it is the logical successor to that of Frobel; insofar as Gerland failed to be completely consistent, Leighly has appended the necessary modifications.

To be sure, Gerland himself did not admit that he was opposing "logic" to "tradition," but claimed that he had given ample regard to the historical development of geography. But the only evidences for that claim are scattered instances where he reaches back to a few predecessors to find, in part of their work, that to which he would limit the whole of geography. His explanation that he "considered only the main course of the development, not the momentary conditions that are of no historical value," reads somewhat ironically today.

Focus on Garland's Thesis : We need not consider Gerland's thesis in detail, but may focus our interest on the two major premises on which it is based. In one he states the nature of a science, in the other the nature of the earth which geography's name requires it to study. The greater part of his lengthy treatment is an attempt to fit these two premises together logically into a "science of geography."

It is characteristic of this type of approach to the methodology of geography that the premise regarding the nature of a science is presented in a categorical statement devoid of any supporting discussion or references. It seems unnecessary to repeat his somewhat lengthy definition, since students of such opposite viewpoints as Wagner and Leighly both recognize it as the description of "a physical science." If this premise be accepted, Wagner agreed that much that follows must logically be granted; but he did not feel compelled to accept a statement of science that would grant scientific standing only to the so-called exact physical sciences. If this major premise is not accepted, the entire subsequent discussion falls to the ground.

One sees at once the disadvantage of attempting to construct the concept of a particular field of knowledge on the basis of a general definition of science. The question of the nature of science is a philosophical rather than a scientific question; that scientists, as such, are not qualified to answer it is reflected in the wide' range of answers that different scientists provide. Great as may be the disagreement among geo-graphers in their views in regard to the proper scope of geography, the differences on this point are relatively small in comparison with those that appear when they attempt to state what science is in general. Furthermore, the viewpoint on this question changes much more radically in course of time. This is reflected in the comparison of Frobel's program and Gerland's. As Leighly correctly notes: in both cases, "it is the voice of nineteenth-century natural science that finds utterance". But whereas in Frobel's time, "natural science" was thought to include the objective study of man, fifty years later the concept of "natural" had been narrowed to exclude man. Further, as Leighly indicates in unconscious agreement with Wagner, Gerland's definition of science is even more narrowly conceived: the basis of his argument is "the strict logic of physical science." It therefore require fixed conclusions and laws of certainties ; a human geography that could only deal with "probabilities" was no science at all Naive as this idea may seem to it today in view of the change in thought forced upon physicists and chemists by the discovery of irreducable uncertainties in the study of electrons, in Gerland's time it no doubt seemed a fair assumption for physical science.

In his second basic assumption, Gerland deduces the field of geography from its name; whether "geography" or *"Erdkunde"* is immaterial, it is the' science of the earth. (This etymological conclusion is also supported by reference to the concepts-in contrast, we may note, to the work—of Kant, Ritter, and one or two others.) The essential question then is: what is the earth? This question, he evidently felt, is not to be answered in terms of some theory—i.e., the earth as an "organism"—nor in terms of a particular, anthropocentric point of view—i.e., the earth as the home of man. Looking at the earth from an objective, scientific point of view Gerland found that it was like millions of other units in the universe, a great complex of materials in state of change, bound together and interrelated by various forces to form a unit Whole, though acted upon also by external forces, notably from the sun. The problem and object of geography is therefore to study the interrelations of these forces and the resultant changes in earth materials. Vast though the problem is, it forms a complete unit.

We need not examine all the logical difficulties into which Gerland was brought by the incompatibility of his two major premises. Wagner discusses these in great detail, though without clearly labelling the fundamental causes. Briefly, the premises are incompatible from two major points of view-one in regard to the content of "earth science," the other in regard to methods of studying it.

What is the content of geography according to Gerland's definition of the earth as its object of study? If the earth materials that are in a state of interrelated development include rock, water, and air, can they exclude these materials transformed into plants? Obviously the' component forces effecting changes in rocks, water, and air include the life force of plants and animals. Equally obviously, the materials of which human animals are made, and those which they transform into cultural products, are earth materials, and the human energy that has altered the face of the earth over great areas is certainly one of the forces affecting changes in earth materials. Indeed it would seem a curious reversal to earlier religious viewpoints to argue that man and human energy were not of the earth, earthy. If, however, as Gerland stated in this premise, all the materials and forces ot the earth represent a unit whole *(einheitliches Ganzes)*, we cannot exclude some of these materials and forces and still have a unit whole left".

The reasoning of the preceding paragraph was essentially that followed by Frobel in explaining the inclusion of man in his "scientific geography." Though Gerland recognized its applicability to plants and animals, and for that reason in part included them in his more limited science of the earth:' he could not follow it through logically because it comes into conflict with his other major premise, that of the nature of a science. Human phenomena could not be included in a science as he had defined it and must therefore be thrown out of geography. In other words, reality must be distorted to fit science, science is not to be constructed to fit reality. This viewpoint is illustrated, almost at: the same time as this is being written, by an English radio song:

"How does a hen know the size of an egg

With no egg-cup beside her

With no one there to guide her?"

Gerland's solution of the logical dilemma was a compromise. Human phenomena are sacrificed to one premise, but non-human organisms, more amenable to "scientific Laws," are included on the basis of the other. That this was inconsistent with his arguments for excluding man was all too obvious: the method of the biological sciences is not the same as that of the physical sciences. (That the physical sciences may all be included under one "method," as he implied, is another undemonstrated assumption that might be questioned.)

Garland's Contemporaries : To Gerland's contemporaries it was therefore clear that to fit the earth to his concept of a science he must change his concept of the earth by stripping it of its plant and animal cover. A further amputation was also required. In contrast with Frobel, who had wish_ to relegate regional studies to a very loosely defined "applied," but "unscientific" geography, Gerland included the study of areal combinations of phenomena (other then human); *Landerkunde* was one of the three ,major parts of his 'geography. Wagner, and later Hettner, observed that such studies could not logically be included under the strict definition of science assumed

Since it was clear to those who read Gerland's thesis that to be logical, he must exclude plant and animal geography as well as regional geography, few geographers considered it necessary to examine his thesis further. That is' to say, they were not ready to throw overboard as surplus baggage not only all the work of Ritter and his followers but likewise the greater part of the work of Humboldt-in general, by far the greater part of the geographical literature of the past, not to mention the scientific organizations and periodicals, and the training of the current generation of geographers. In his historical study, therefore, Hettner properly limited his consideration of the thesis to its lack of continuity with the historical growth of the field; its logical claims however he did consider in his first full methodological treatment, in 1905. Likewise Penck attacked the logic of the thesis in an address given in this country, at the Congress of Arts and Sciences at St. Louis in 1904.

We may say, therefore, that logically Gerland presented not one thesis but two conflicting theses. If one follows consistently his definition of the earth, then geography, as the study of the earth, would include all that had commonly been studied in geography and also all of geophysics if not also of geology. It would not however be a science as he defined that term. If one follows consistently his other premise, as Leighly does, one may prove that, in the terms of "the strict logic of physical science," geography must be a physical science of the earth, i.e., "geophysics." But one may well wonder why it required over 50 pages and 50 years to prove that proposition.

The historical fact is that Gerland's contemporaries saw clearly enough the ultimate conclusions to which either of these alternatives led. Although they recognized that the methodological statements and general outlines of earlier geographers had commonly included the study of the earth body, they were sufficiently realistic to recognize that the actual contributions of geographers to geophysics had been extremely slight and that "the knowledge and training inherited in geography did not equip its students to work in that field. Although a few still continued to include the study of the earth body in their definitions of geography—Richthofen, in particular, returning to this concept in his Rectoral address at Berlin, in 1903—in practice, the geographers in Germany as well as everywhere else, confined themselves to the study of the earth surface. To be sure, they recognized that a complete explanation of surface phenomena-at least of land forms—must ultimately be based on deductions concerning the interior, but they did not suppose that one branch of science must include all upon which it depends. As Bucher had observed in that connection, every science depends in part on other "help-sciences," and any science may serve as a "help-science" without thereby losing its independence. Properly speaking, no branch of science can be a complete unit in itself independent of other branches.

Wagner's View : In reality, as Wagner explained in detail, the examples of geophysics presented in the volume of studies of which Gerland's thesis was an introduction, gave little promise that studies of the earth's interior could form the basis of detailed studies of the surface forms, or vice versa. Logically, as Gerland said, the study of relief forms should begin with an explanation of the phenomenon of the continents, but after a half century of work in geophysics what "certain" explanation do we have for the "particular form of the continents?".

Even more clearly did Gerland's contemporaries foresee the consequences that would result if geography were transformed into geophysics. As 'followers of Humboldt they saw the particular value which geography offered to all science in forming a connection between the artificially separated portions of reality studied by the natural and the social sciences. They did not share Gerland's confidence that the social sciences could be trusted to make these connections themselves, that geography need only provide the raw material, in the form of knowledge of the physical facts arranged by categories, or-in Gerland's original thesis-combined within particular areas in their relations independent of man.

Specifically, Wagner asked, if the geographer is to "present scientifically," for the benefit of "the sciences to which geography is auxiliary," "the nature and natural productivity of a country, that interest the geographer for their own sake" what is this "natural productivity i.e., productivity independent of man (and according to the improved form of Gerland's thesis, independent of the vegetative cover). This question has been raised in various symposia of American geographers; many readers may remember a discussion concerning the possibility of a "pure geography" of the Amazon Basin. Since the same combination of soil and climate may be highly productive for some crops and of low productivity for others, the very concept of "natural productivity" has no meaning excepting in reference to the particular crops that particular men wish to grow. We could map the natural productivity of an area only by making separate maps for every conceivable combination of agricultural crops and methods.

Further, various of Gerland's critics objected that even if we eliminate all these difficulties and reduce geography to the study of the physical aspects of the earth, literally in terms of the physics of the earth, such a science would have no more unity than the science of physics itself. It would represent, that is, not a part of physics, but a collection of all the parts of physics that can be applied to one particular object, the earth. This is most clearly shown by Leighly's reference to the several sections of the American Geophysical Union-an organization whose very title indicates that it does not represent a branch of science, but rather a collection of branches of science that have certain interrelations. If we add together meteorology, seismology, vulcanism, terrestrial magnetism, geodesy, and hydrology we have a sum that represents much of the field of physics as applied to various physical phenomena of the earth. It is united only as an applied field. In terms of pure science it represents widely separated branches of physics in which the kinds of phenomena studied and the methods necessary for studying them are as heterogeneous as could be found in all of physics. The sum of these branches does not form "a physical earth science" but rather many physical earth sciences—only one of which is commonly called "geophysics"—and if we attempt to put these many together we return to the field of physics as a whole.

Likewise, finally, Gerland's critics saw that, in trying to construct a logical edifice for the field of geography on the basis of a particular concept of science, Gerland in reality had not considered the problem from the point of view of science in general. He recognized that there were interrelations between human and non-human phenomena that should be studied in certain fields of science but at the same time denied scientific character to these fields. Any field that studies the interrelations between these heterogeneous phenomena must use both of the "methods" that he found incompatible within a single science. Why should the social sciences have this duality of method imposed upon them, or on the other hand, if they must and call employ both methods, why may not geography?

For example, Gerland vigorously objected to Ratzel's attempt to develop a geographic law concerning the separation of peoples by mountains: "the appropriate law here is not found in the connection of men with mountains, but in certain psychophysical reactions that are released by the mountains, hence does not lie in the field of geographic thought, but completely in psychology and physiology.

Gerland is more consistent, however, in other parts of his discussion, indicating that such relationships are not to be studied in science at all-in what other form of knowledge they are to be studied he does not say. "A science of the heterogeneous things that fill area is impossible because of their heterogeneity". The significance of this statement is not entirely clear until we come to his illustration of it, almost at the end of the essay: "What sort of a connection is involved in the fact that Rome is located on the Tiber, Prague on the Moldau? River and City are heterogeneous concepts that geography can never logically unite." Since neither the particular river Moldau is logically necessary to the particular city Prague, nor vice versa. "an intrinsic connection does not exist".

The answer of geographers since Herodotus and Strabo—including Humboldt and Ritter and the followers of both is so obvious as to appear almost superfluous in constructing the city of Prague man has produced a synthesis of city streets and buildings with the particular character of the river Moldau at that site. Though, in this case, the economic and political centre of the Bohemian basin might perhaps just as well have developed on one of the other headwaters of the Upper Elhe, such a city-whether called Prague or not-would have been significantly different from that which actually exists. Further, if we look beyond the particular examples that Gerland selected as apparently suitable to his thesis, would he have maintained that the particular river Rhine is not essential to the city Cologne, that there is no significant relation between the city of New York and New York Bay?

Specific Illustration of Garland : Gerland's specific illustration of his thesis brings us clearly to a parting of the ways. Those who follow him to the logical conclusion will study the phenomena of the earth in clearly separated categories; they will not, like Ratzel, concern themselves with the contrasts in population in successive zones of elevation in Scandinavia nor with the zones of land use at different elevations in the Alps. Neither win they, like Humboldt, study the differences in natural vegetation at successive zones of elevation in the Andes, nor in the different climatic zones of the world. On the other hand is found that host of students who from earliest times have observed that these heterogeneous phenomena exist in reality not merely beside each other, but in causal relation with one another, and they have devoted themselves to the study of areas of the earth filled by these things in their interrelations. This study, since the days of antiquity, they have called geography and the world of knowledge has recognized it under that name. Whether or not the study is a science is for them a secondary question, it is the field of study that has long since been claimed by, and acknowledged to, geography.

We arrive therefore at the ultimate conclusion—obvious enough, one may say, from the start to which "the strict logic of physical science" leads us. Granted the need for a study of the physics of the earth, we need only recognize how completely different such a field is from that which scientists as well as laymen have for centuries called geography. Gerland's thesis, as completed in the discussions then and since, may well stand as the *reductio ad absurdum* of all attempts to make geography into either an "exact" or an "essentially natural" science.

Earth's Science

There is one further problem, on which Gerland's essay throws light—namely the question whether the field of geography includes the earth as a whole as its name certainly implies, or whether it is limited to that part of it which we know directly-the earth surface. In most methodological discussions prior to Richthofen's, the former was assumed, and most text books opened with a section on the earth as a body-as is true of some recent texts. On the other hand, the actual studies of all but a very small number of geographers were limited to phenomena found within the thin shell of the earth surface. Geography, in practice, therefore was the study of the world, as that term is most commonly understood.

Unquestionably however it seemed arbitrary to attempt to restrict a science called the study of the earth to the outer shell of the earth. Richthofen, recognizing that the practice of geographers had in fact defined their field in this way, had attempted to defend the restriction logically in his inaugural address at Leipzig, in 1883. Although Penck found this "never to be forgotten address" completely convincing, other students, including Hettner, felt that Richthofen had not found the necessary logical basis for the thesis, and Richthofen himself departed from it later.

Gerland's discussion of Richthofen's address does not help us; a certain lack of clarity in the original argument permitted Gerland to misconstrue it into the opposite of its intention. But in his own manner of viewing the earth as the object of study in geography, Gerland provided, to my mind, the key to the fundamental distinction in point of view between geography and geophysics. Although this presents a problem to few geo-graphers today, there is evidently sufficient uncertainty on it to justify a brief consideration.

In order to give an objective, scientific answer to the fundamental question, "what is the earth," Gerland discarded not only the teleological and anthropocentric views, but also, by implication, the geocentric viewpoint. Looking at the earth from a cosmic point of view he saw it as but one of many similar units in the universe. There can be no question that from this point of view, the earth body would be included in an earth science, indeed the interior of the earth would appear vastly to outweigh the surface. One might then conclude that the actual concentration of our study on the surface is simply a result of our inability to make direct observations on the interior, and of the fact that the surface provides the expression of the changes in interior materials produced by interior, surface, and external forces.

While this point of view is logically sound, it is fair to ask whether it is as realistic and objective as it appears. Unless we are to determine the relative importance of different things in the universe solely on the basis of mass, it is just as arbitrary to consider the earth as no more' than one of the smaller planets of one of the lesser of millions of stars in one of many galaxies, as it is to describe it as the home of man-both statements are equally true. The very fact that we have a science of geography, but no science of Marsography reflects the two objective facts that make the earth, for us, unique, and justify our distinguishing it out of all the units of the astronomer's universe for study in one or more earth-sciences.

One of these facts and literally speaking we cannot get away from it is that the earth is where we live. We may say that science is not in itself an abstract thing, but is rather man's pursuit of knowledge of his universe. In this universe the earth is, for science, unique above all other units, as the one on which man lives. If other satellites of our sun, or of any other, were known to be like ours in all important respects, we would no doubt have a special science concerned with studying them, but would at the same time require a separate science for the separate unit on which we not only live, but rather of which we are a part. In the comparative science of earth-like planets we would look at the earth from a more celestial point of view; but from that point of view there would be little justification for a separate science of the earth. The justification for that science would rest only upon a geocentric view.

The previous consideration however is, in the fact, academic. In our actual science of the cosmos-however detached our point of view-the earth is unique. Whatever speculations one may make as to the possibility of the existence-of similar planets-recently calculated by Jeans as a very minute possibility-our science knows no other planet like it. In other words, for a celestial scientist, attached to no planet or star, but knowing only what we know, this little planet Earth would be an object of unique and extraordinary importance. It is the one body in the universe on whose surface temperatures fall within that very narrow range in which different materials exist simultaneously in solid, liquid, and gaseous state; in consequence of these and other factors it is the body whose surface shows the greatest multiplicity of varying interactions of factors and differential results; further, on it alone can that extraordinary phenomenon called life be observed in millions of generic and individual variations; and finally, to our celestial scientist it would not be a matter of indifference, that on this planet alone (so far as science can say) creatures capable of considering science can exist. Indeed, to become real our celestial scientist must come to earth-only on the earth can he find his home.

The reader will have observed in the previous paragraph that the uniqueness of the planet earth to science is expressed, entirely, in terms of conditions on or near its surface. Our unique interest in the earth is a unique interest in the earth surface. If it be argued later that the understanding of the surface requires a knowledge of the interior, it may be necessary to dig into the interior-so far as that is even intellectually possible. But the object of our study in a science devoted to the earth as a unique planet, is the earth surface—the thin shell of outer earth in which are found the unique phenomena of earth. In other words, the geocentric point of view, which is implicit in the existence of a science devoted exclusively to the tiny fraction of the universe called the earth, is centered not on the astronomical, planetary unit, but on that naively given section ,of reality that man, from earliest times, has thought of as the world of which he is a part, and which only recently he has discovered to be but the outer shell of an astronomical body which he has since called the earth.

Change in Knowledge

The change in our knowledge of, the earth, referred to in the preceding sentence, may well be the explanation for the fact that we have no single word of common speech that refers unquestionably to the field commonly ascribed to geography, thus forcing us to use a term that appears artificial and at the same time not clearly defined-namely "earth surface." On the other hand, as Lehmann has recently pointed out, every one-outside of tour circles at least-understands what we mean by the word "world particularly in the many combinations in which it is commonly used: "world tour" "world map," "world trade," "worldwide," "World-War," etc. In all of these cases there is no uncertainty as to the physical extent of space included : they do not include the centre of the earth, nor do they include the moon or other "heavenly" bodies, but simply that outer shell of our planet as high in the atmosphere and as deep in the ground as man can *experience.*

If the physical scope of geography is considered as simply the world as it has almost always been in practice—it seems unnecessary to attempt any sharp definition of the term "earth surface." Biologists have little difficulty in studying the field of "living bodies" even though they may find it extremely difficult to give an accurate definition of "life." Likewise geographers need not concern themselves with establishing an exact boundary between the "world" and the interior of the earth. If the explanation of surface phenomena requires deductions concerning conditions deeper in the interior, the geographer who is able to dig deeper will presumably not stop at any lower boundary that might be set; so long as he maintains the purpose of his study in the surface-the world-there is no need to set such a boundary.

If the earth surface be regarded as the domain of geographic study, forming in its entirety for geography alone the sole concern, it also forms at the same time, as Richthofen indicated, the meeting place for all sciences that are concerned with the earth: meteorology, geology, geophysics (in the narrowest sense), geography, economics, etc. But it does not follow that each of the other sciences enters this area as the guest of geography as Gerland's construction of one unfortunate expression of Richthofen's had him say-but rather that each of these other sciences is, with geography, equally at home in some part of the earth surface. But each of these is interested in the earth surface from a different point of view, just as mineralogy and paleontology may study the same rocks from entirely distinct paints of view. Richthofen failed to make clear this distinction in point of view; apparently like all the geo-graphers of his time he overlooked Humboldt's statements of the contrast between the geographic viewpoint and that of the systematic sciences concerned with the same objects. By the time this view was again clearly expressed, by Hettner in 1905, the particular argument concerning the place of the earth-body in the' field of geography had become of little more than academic interest. Wagner's prediction that the highly specialized techniques required in geophysics would prevent geographers from making important' contributions in that study was apparently fulfilled and geophysics became definitely established as an independent field in close relation to physics and geology.

Relationship Science

To many readers the outline of historical development of geographic thought may appear to give too slight attention to what was for half century the dominating concept in American geography, namely the study relationships between nature and man. In view of the importance of concept, not only among research workers in geography in this country and in England, but particularly in the schools, it is necessary to give it special attention.

Throughout the previous discussion we noted that geographers seriously interested in developing their subject as an independent branch of science have constantly sought to understand the complex relationships existing among the vegetation, fruits, occupants, and peculiarities of the various quarters of the earth-to paraphrase the statement quoted from Strabo. In any science the study of its phenomena involves the study of relationships that may be found among them. In geography, in particular, such relationships form the connecting links between heterogeneous and otherwise unrelated phenomena, uniting them in the single study. But we give an entirely different orientation to a field if we focus the attention on the connecting links, rather than on the sum total of phenomena interconnected in areas.

We have seen that the idea of defining geography in terms of relationships developed only when the subject as a whole was thrown farthest out of balance-in the latter part of the last century-in order to retain the study of human phenomena in a field in which, according to its particular development at the time, they seemed out of place. With the restoration of a balanced view later in the century the concept gradually dropped out and is hardly to be found in methodological discussions in Germany in the past two decades.

It is an extraordinary fact that, while modern geography in all countries has been fundamentally dependent on that developed in Germany, the particular concept which influenced German thought for but a brief period, should have come to dominate the methodology, and much of the practice of geographers in France, England, America, and apparently Japan. Though simple explanations of cultural developments are to be included among the less reliable of historical conclusions, this particular result appears to have been caused by the great interest which a small number of outstanding foreign students took in the anthropogeography of Friedrich Ratzel. Although this work was of great importance in establishing systematic (general) human geography, Ratzel's methodology-whether in his works or in specific methodological statements-does not appear to have had any important effect on the methodology of present-day geography in assumptions.

Though this statement, which is in marked contrast to the assumptions of most English-speaking geographers, is not expressed directly by German writers, it may be inferred from the small attention paid to Ratzel's works in methodological discussions, and has been substantiated by personal statements. On the other hand, it should be added, parts of Ratzel's work of which Semple gave but little indication have also been important in German geography, His regional studies have been highly regarded, particularly by those interested in the aesthetic aspects of the subject, Relatively speaking, his importance was greatest in political geography, A sociologist, L, Gumplovicz, says: .. Ratzel's works contain more, and more important, knowledge concerning the state, than the entire theoretical political science literature of the last hundred years". Granted that there may be some exaggeration in that statement, Ratzel is universally" recognized as having established the foundations of this particular branch of modern geography,

The close association, in France, of geography and history may account for the active interest in Ratzel's work taken by Vidal de la Blache, Vallaux, and Brunhes, At least one article by Ratzel was published in France, in French, and his works were given lengthy abstracts and reviews, so that, for a considerable time, French geographers were greatly concerned to criticize, modify, and correct his statements and point of view, As similar consideration was not paid to other German geographers, then or later, Brunhes, in a short outline of geography, published as late as 1925, gives more than four pages to Ratzel, less than half a page to Richthofen, and dismisses later German geographers as amplifiers and modifier of Ratzel's work. As Sauer correctly observes, Richthofen's influence has been far more important than Ratzel's in twentieth century German geography.

Brunhes' methodological considerations are perhaps more suggestive than definitive but his detailed portrayal of his concepts, in *La geographie humaine,* presents essentially a modification of the dualistic concept of geography, in which a large part of human geography is defined in terms or relationships. The most thorough methodological study in French geography, and perhaps the ablest presentation of the concept of human geography as the study of relationships (in which however physical geography is defined in terms of distribution) is Vallaux' *Les sciences geographiques*. On the other hand, these works have been less important in determining the direction of French geography, than the concept of regional geography which Vidal and his school have developed to such a notable extent, particularly in monographic studies of particular regions of France.

Ratzel's most important disciple was an American, Ellen C. Semple. By her rendition of his ideas in her teaching, and in her publications, she had a marked influence not only in this country but also in Great Britain—where Chisholm's enthusiastic discussion of Hettner's methodology appears to have passed unnoticed and Herbertson's ideas of regional geography long met with opposition or indifference. In this country the association of Semple's concept of "geographic influences" with the frankly dualistic concept which Davis had brought from Germany largely determined the methodological thought of at least a generation. It was expressed repeatedly in the presidential address of this Association from 1906 to 1926--with the notable exception of that of Fenneman—and is most fully developed in the works of Semple, and Whitbeck and Thomas. Dominant in the universities until recently, it is still the prevailing concept in the teachers colleges and secondary schools. It is therefore necessary for us to examine this "environmentalist concept" in some detail, even though the objections to it have been stated repeatedly in American' publications during the past dozen years.

Until comparatively recently the environmentalist concept was commonly expressed in terms of the influences of natural conditions on human inhabitants, or human activities. The natural environment was studied in itself, with little question as to its limits; indeed it was suggested that was "pure geography," and many geographers and, even more, social scientists came to speak of the natural environment as "the geographic factor" human geography was the study of the influence of this factor on man. Geography was therefore dualistic not merely in the sense that it studied both natural and human phenomena, that it must use techniques of various natural sciences and of various social sciences—no science can limit itself to its own particular methods—but rather in the sense that its fundamental point of view was different in studying natural and human phenomena. "Dualistic" therefore, Hettner suggested in 1905, implied the other meaning of its literal German translation, *"swiespalting,"* that is, discordant. The "logical unitary structure of geography was thereby destroyed," whereas "if the conception of nature and man proceeds from the chorological viewpoint, it is homogeneous in all important points".

In a lengthier discussion, published in *1907*, Hettner noted that it was unsound to proceed from the consideration of natural factors to the explanation of human phenomena. The consideration of the natural factors can arrive only at possibilities; the decision lies with man. . . . A sure scientific knowledge of the actually existing causal relationships is only possible if one proceeds from the human facts, classifies these and pursues them to their geographic roots". Much the same argument is to be found in Barrows" presidential address of 1922. He recommends that relationships in geography should be studied "from man's adjustment to environment, rather than the reverse. The former approach is more likely to result in the recognition of all factors involved to minimize the danger of assigning to the environmental factors a determinative influence which they do not exert".

Barrows' modification of the environmentalist concept of geography has apparently the further effect of solving the problem of "dualism." But it does this by the elimination of the study of physical geography in itself. Both natural features and human features are studied in geography only in their relations to each other. Geography is therefore exclusively human geography, or as Barrows stated it, "human ecology;" it is therefore essentially a social science related to the natural sciences in the same manner that plant ecology as a biological science is related to the physical sciences. In theory, therefore, this concept represented a departure from the previous history of geography which was very nearly the opposite pole from that suggested by Gerland. Its logical conclusions have been most definitely expressed by a group of geographers in Japan who hold that physical geography is no longer a part of geography, but is to be found in other sciences, that what remains to geography is exclusively human geography. In practice, of course, there can be no clear separation, since a study of the adjustments of man to the environment requires a knowledge of the environment, but this knowledge' is logically subordinate, not to be studied for its own sake.

On closer consideration of this concept of geography as human geography, its apparent unity appears much less real. If considered, as formerly, in terms of geographic influences on man, geography, as anthropogeography was a collection of relationships which could be classified in categories according to each of the natural factors, but in each of these collections there was no basis for a unified organization of all sorts of relationships, and the collections together had likewise no unity. It is no wonder therefore that undergraduate students complained of the encyclopedic character of Semple's major work; probably most research students find it of value chiefly, if not solely, as a reference volume. In other studies in which Semple succeeded in establishing an organized unity, the organization will be found to be primarily historical, but on the other hand, from the historical point of view, the work is only partial.

In a sense, therefore, Semple had in practice used the reverse form of the environmentalist concept-the relation of man to nature-in her study of *American History and its Geographic Influences*. The organization therefore is in terms of the social sciences, and geography on this basis has in one sense the same degree of unity as the whole of the social sciences, but in another far less: it organizes its collection of relationships in terms of the organization of the social sciences, but within each of these it presents but a part of the total. Dickinson has recently noted the dispersive effects of this view of geography in the work of British geographers

A further modification of the definition of geography, which describes the field as the study of relationships, or adjustments, within regions, provides an easier organization of the material, no doubt, but one which appears purely arbitrary. If the major concern of geography is with relationships, presumably the logical organization would be in terms of the character of the relationships; the particular places in which they were found, or the particular times in history in which they were found, would be irrelevant. In any case, even though the regional classification provided a convenient method of restricting the view, it would provide no basis for a comprehensive organization of the relationships found within the region.

Environmental Concept

Although the environmental concept, therefore, appears to provide no logical basis for unifying geography internally, we must consider whether it does not at least set off geography as a distinct unit from other fields. We have seen that Hettner stated, as early as 1907, the conclusion which many "environmentalists" in this country adopted in the 1920's—namely, that the procedure of study of relationships should lead from man to nature, rather than the reverse. Hettner however followed the argument 'farther, by asking how a science can come to "arrive at an answer to the question of the dependence of man on the nature of the earth. The function of answering this question cannot be the domain of one science. The systematic sciences that study man must establish connections with the nature of the earth surface if they are to understand the different social conditions of different lands and places. Historical writing must take account of the natural conditions if it wishes to explain and understand individual events, *e.g.*, the course of a campaign, the territorial changes of a peace treaty. . . etc., or if it wishes to explain and understand the whole course of man's historical development. Geographers are not to be jealous of such studies of other sciences; on the contrary, we should rejoice in them, accept their conclusions thankfully and come gladly to assist in them. But in such assistance, our considerations must retain their centre of gravity in geography".

It can hardly be denied that many geographers have regarded as trespassers those students in other fields who presumed to study relationships between man and nature. Even more have students of other fields wondered whether geography had any field of its own. Defined not in terms of a particular circle of facts, but in terms of causal relationships presumed to exist, it could have but a parasitic character.

If we depart from theory for the moment and examine the character of the work produced in geography under this concept of its nature, we must recognize that there is ample evidence to support the accusation of our colleagues from other fields. Nc one, to be sure would wish to lay down fixed boundaries patrolled by border guards between adjacent sciences, and it is both natural and proper that some of the workers in any field wilt be engaged near the borders and frequently overlap into a neighbouring field, persuadably to the benefit of all concerned. Further, geography unquestionably has close contacts with a large number of other fields. Nevertheless, if geography is an independent field of study, it would seem a fair assumption that most of its workers, most of the time, would be found working' within that field. That appears to have been in fact the case among German geographers of the past generation (if we can overlook a possible' argument in regard to geomorphology), and, in large part, among the followers of Vidal in France. In contrast, the geographers whose work has most clearly been governed by the concept of relationships have almost all carried on a large part of their studies in areas that must at least be regarded as transition zones, if not definitely parts of other fields. Semple, Brunhes, Huntington, and Taylor-to mention only the most prominent—have all published major works that were frankly indicated as combinations of geography and history, and each of the last two named have also explored in various other' fields. If any should suppose that these are exceptions, he needs only to thumb through the past volumes of these Annals and note the large number of smaller studies that are as much studies in other fields as in geography. The attitude of students in the other fields towards these studies—whether, as in many cases, they find in them new ideas and new methods useful in their fields, or whether they find them lacking in the necessary background of their fields—need not concern us here. What doės Concern us, is that such studies offer no answer to the question as to how geography may be distinguished from other fields of science.

An even more unfortunate result of this situation has been that students in other social sciences have been brought to look upon geography as essentially an a Priori thesis which, in the extreme form, is simply environmental determinism-the mechanistic form of predestinarianism. Overlooking the unfair attacks of many critics—based frequently on work by non-geographers, as Whitbeck and Thomas rightly note there is much that cannot be refuted in the detailed and critical, but by no means unfriendly; examination of geographical work by Febvre. This unfortunate result is not to be eliminated merely by inserting qualifying terms, by speaking of "possibilities," etc. As Schluter, Michotte, and Sauer have all stated, a definition of a field of science in terms of causal relationships instructs its student to seek and find such relationships, robs him of his impartiality, and easily leads to dogma; for "success lies most apparently or at least most easily, in the demonstration of an environmental adjustment". If it were necessary to demonstrate this proposition Our geographic literature would offer all too many examples. The most extreme forms tend to be promulgated just where the greatest reserve is desirable, in school geography; I have listed a number of examples of these elsewhere.

Finally, the concept of geography as a study of relationships has failed to provide the subject "with sufficient tangible objectives," that is, with concrete phenomena to be studied, or with "a distinctive and sufficient method or discipline"

Because of these logical and practical difficulties, Hettner and Schluter, and following them Michotte, Sauer, and many others, have concluded that "an independent science can never have for its object merely causal relations, but must rather apply itself to a particular circle of facts which it first establishes and describes in order then to search for causal relationships".

Distribution Science

As early as the pre-classical period of modern geography, it was suggested that geography is essentially the science of distributions, a field whose purpose is to study the distribution of different phenomena, separately and in relation to each other, over the earth. Doring found that such studies represented a minor part of the geography of Humboldt. This view of the field was presented with greatest emphasis by Marthe, who described geography simply as the study of "the Where of things". More recently it has been represented by De Geer, who in other respects parallels Hettner. The latter, Hettner, was at one time accused by Schluter of supporting this concept, but it was only necessary for Hettner to repeat his previous arguments (to be cited later) in order to demonstrate that he had been totally misunderstood.

If the study of distributions is fundamental to the character of geography, not merely incidental to other purposes, it must form a characteristic distinguishing geography from other fields. This necessary conclusion has led many to suppose that when a botanist determines the areal distribution of a 'particular plant, or a geologist determines the location of a volcano, or a sociologist maps the distribution of population in a country, instead of merely using a statistical table, the botanist, geologist, or sociologist becomes thereby a geographer or, at least, is working in geography. But, as Michotte concludes from this series of examples, each of these students is carrying on work necessary to the understanding of the particular kind of phenomena that he is studying from the point of view, not of geography, but of his own science. The use of the inductive method, in any of the sciences dealing with phenomena located within the earth surface, in the effort to establish the relationships that govern the character and development of the phenomena of that particular science, will often require the determination of the location of the phenomena before any principles can be determined.

Beginning with his earliest methodological discussion, in 1895, Hettner has repeatedly stated similar objections to this concept of geography. "Distribution by place," he wrote in 1905, "forms a characteristic of the objects . . . and must therefore necessarily be included by the systematic sciences in the compass of their research and presentation." He emphasized the distinction that Wallace had made between a geographical zoology that studies' the distribution of the individual genera and species of animals, and a zoological geography-or simply animal geography-that is concerned with the difference in faunal equipment of the different lands. (Michotte notes the importance of this distinction in titles. The same distinction in point of view, Hettner illustrated by several other examples of natural features. In each case, the view of the systematic science is focussed on the phenomena, which are studied in their distribution, that of geography on the *areas* that differ from each other in their mineral, floral, or faunal contents. This geographic view represents, we may note, a consistent derivation from Humboldt's description of geography as the study of "that which exists together in area". It appears most clearly, perhaps, in Hettner's example with regard to cultural phenomena: "A study of the distribution of a particular tool, weapon, or in general of any particular object or custom is erroneously called anthropogeographic; it is much more ethnological, even though it may indirectly acquire anthropogeographical significance; for the point of interest here is not the land, but the object concerned or the people as possessors and bearers of this object".

The comparison of geography and history, upon which Hettner bases his general concept of geography to be considered in full in the following section, might suggest, as Sauer notes, that if history may be considered as the science of the When, geography would be logically the science of the Where. Neither Hettner, nor Michotte, who accepts Hettner's comparison of the two fields, fall into this error. Though "When" is one of the several questions that historians always seek to answer, just as the geographer is always concerned to know where his phenomena are, historians do not regard their field as defined simply in terms of the element of time. Not because of the limitation suggested by Sauer, which they do not accept, but rather, as Kroeber puts it, because history, as a distinctive field of knowledge, is not concerned with analyzing the processes of particular kinds of phenomena-leaving that to the systematic sciences—but rather with the integration of different kinds of phenomena with respect to periods of time.

Finally we may add the very important practical objections that Sauer raises against the concept of geography as the science of distributions. Geographers would be influenced to work "on the distributional expression of many facts wherein our competence is less than that of many others further, "there is known no discipline for the development of such studies;" and, finally, "distribution per se does not supply a common bond of interest".

Even though most geographers today would accept the conclusion that the special function of geography among the sciences is not to be found in the study of distributions, it is extremely difficult to avoid the apparently logical conclusion-often held unconsciously that because the geographer cannot study any phenomena without knowing where they are, the study of the Where is in itself a part, at least, of our field. As this is a current problem, primarily in systematic geography, we will need to consider it again in examining that aspect of the field. For the present it is sufficient to note that all of the systematic sciences that are concerned with phenomena located within the earth surface must also answer the same question in at least part of their work. The problem of determining and explaining the Where of things is not the distinctive function of geography.

10

The Morphometry

Meaning and Nature

The study of the characteristics of drainage network of a particular region is approached in two ways e.g. (1) descriptive approach and (2) genetic approach. The descriptive approach involves the study of the characteristics of the forms and patterns of the streams of a given region while the genetic approach involves the investigation of the evolution of streams of a region in relation to tectonics, lithologies and structures. Thus, drainage system refers to the origin and development of streams through time while drainage pattern means spatial arrangement and form of drainage system in terms of geometrical shapes in the areas of different rock types, geological structure, climatic conditions and denudational history e.g., trellis pattern, dendritic pattern, parallel pattern etc. The examples of drainage systems are consequent streams, subsequent streams, obsequent streams, resequent streams, antecedent and superimposed streams etc.

The origin and subsequent evolution of any drainage system in a region are determined and controlled by two main factors viz. (1) nature of initial surface and slope and (2) geological structure (e.g. folds, faults, fractures, joints, dips and strikes of rock beds and types of rocks). Streams or drainage systems are divided in two broad categories on the basis of the adjustment of the streams to the initial surface and geological structures e.g. (1) sequent streams (which follow the regional slope and are well adjusted to geological structures) such as consequent streams, subsequent streams, obsequent streams and resequent streams, and (2) insequent streams (which do not follow the regional slope and are not adjusted to geological structures) such as antecedent streams and superimposed streams.

Major Systems of Drainage

Sequent Drainage Systems

Consequent Streams : Consequent streams are the first streams to be originated in a particular region. These streams have their courses in accordance with the initial slope of land surface. In other words, the consequent streams follow the regional slope. These are also called dip streams. In a region of folded structure (when the crustal rocks are folded due to lateral compressive forces into parallel anticlines and synclines) consequent streams are formed in the synclinal troughs. Such consequent streams are called synclinal consequent streams, which become the master consequent streams of trellis drainage pattern at much later date.

The first streams to be initiated on a newly emerged coastal plain are consequent streams which are parallel to each other and thus form parallel drainage pattern. The longest stream of the whole system of consequent streams is called master consequent. Most of the streams draining the coastal plains of India are the examples of consequent streams. The most ideal landscapes for the origin and development of consequent drainage system are domes and volcanic cones. Consequent streams are divided in two types e.g. (1) longitudinal consequent (which follows the axis of the depression or syncline in a folded structure) and (2) lateral consequent (which follows the sides of the depressions or the sides of the anticlines). Lateral consequent streams generally join the master or longitudinal or synclinal consequent more or less at right angle.

Subsequent Streams : The streams originated after the master consequent stream and following the axis of the anticlines or ridges and the strikes of beds are called subsequent streams. Some of the scientists have opined that the subsequent streams originate on the flanks of the anticlines and join the master consequent at almost right angle (which means that the lateral consequents, as referred to above, are the subsequents) while others maintain that the subsequents are parallel to the master consequent. S.W. Wooldridge and R.S. Morgan (1960) have opined that "it should be noted that all the first generation of tributaries to consequent streams are subsequent in the sense that they arise subsequently to the establishment and incision of the consequent streams." According to them "such streams, starting as gullies on the sides of the primary consequent valleys, discover and explore belts of structural weakness, due to softer strata, fault, or joint-planes, and shatter zones" (Wooldridge and Morgan, 1960, pp. 173-74). E. Ahmad (1985) has remarked that "the term subsequent may be used generally not to indicate a chronological sequence but a relation between drainage on the one hand and geology, structure and tectonics on the other. Generally, it refers to the streams transverse to the master consequent." The Asian rivers, a tributary of the Yamuna river and the Song river, a tributary of the Ganga river in the Dehra Dun valley (infilled alluvial plain) are the examples of subsequent streams while the Yamuna and the Ganga are the master consequents.

Obsequent Streams : The streams flowing in opposite direction to the master consequent are called obsequent streams. In fact, obsequent streams are also consequent be cause they also follow the slopes of the ranges. The streams originating from the northern slopes of the west-east stretching ranges of the Himalayas flow northward to meet the east-west flowing tributaries (subsequent streams) of the southward draining master consequent streams. For example, the Mahabharata Range of the Lesser Himalaya has issued several streams from its northern slopes. These northward flowing streams join the subsequent stream Son Kosi which runs west to east, as obsequent stream because these are opposed to the directions of the master consequent like the Ganga and the Yamuna. Similarly, several streams originating from the northern slopes of the Shivalik Range drain due northward to join east-west subsequent streams of the southward flowing master consequent of the Ganga and the Yamuna, north of the Someshwar Range of Dundwa Range (of the Shivalik Range). These northward flowing tributaries are the examples of obsequent streams.

Resequent Streams : The tributary streams flowing in the direction of the master consequent are called resequents. These are originated at much later date in comparison to the master consequent. Since they are of recent origin, and hence they are called resequent.

The resequent streams are originated during the initiation of second cycle of erosion in a folded structure. The gradual denudation of folded mountains during the first cycle of erosion results into inversion of relief with the passage of time wherein anticlinal ridges and synclinal valleys are converted into anticlinal valleys and synclinal ridges respectively. Thus, longitudinal streams are developed in the anticlinal valleys. These features are peneplaned 'by the end of the first cycle of erosion. The initiation of second cycle of erosion begins with the excavation of new valleys in the synclines. Thus, the streams developed in the synclinal portions become resequent streams which are though analogous to the original longitudinal consequent developed during the first cycle of erosion but, in fact, they are many hundreds of metres below the initial surface.

Insequent Drainage System : The streams which do not follow the regional slopes and drain across the geological structures are called insequent or inconsequent streams. Antecedent and superimposed streams are the best representative examples of insequent drainage systems.

Antecedent Drainage System : Antecedent streams are those which are originated prior to the upliftment of land surface. In other words, antecedent streams antedate the upliftment of an upland or mountain across which they have maintained their present courses through continuous downcutting of their valleys. It is, thus, obvious that if a river has developed in a particular region and if the subsequent uplift or upwarping of the land area across the pre-existing river does not divert or deflect the course of the river and the river maintains its previous course through downcutting its valley at the rate equivalent to the rate of uplift of the land area, the river is called antecedent or anti-consequent. It may be pointed out that the concept of antecedence of a particular river is based on guesses and conjectures because neither the rate of the uplift of land area nor the rate of down cutting by the rivers is known.

The nature and the rate of upliftment of land area is very important parameter for the development of antecedent drainage system because rivers cannot maintain their previous courses during all types of upliftment. For example, if the upliftment of the area occurs very rapidly and is completed within a very short period of time, the rivers draining through that area cannot maintain their previous courses because the rate of downcutting of their valleys cannot keep pace with the rate of the upliftment and thus the rivers are dismembered and adopt different courses depending on local (new) conditions. On the other hand, if the rate of upliftment is such that the rivers are able to degrade their valleys through active downcutting (valley incision) at the rate equal to the rate of the upliftment, they become able to maintain their previous courses and thus they become antecedent streams. The nature and dimension of the upliftment also determine the antecedence of the rivers. It is the local and not the regional upliftment which gives birth to the development of antecedent streams.

Examples : Examples of antecedent streams are found in almost all of the folded mountains of the world. Many of the major Himalayan rivers are the examples of antecedent streams e.g., the Indus, the Sutlej, the Ganga, the Ghaghra, the Kali, the Gandak, the Kosi, the Brahmaputra etc. It may be pointed out that the Himalayan mountain system consists of three parallel ranges running from west to east viz. (from north to south) the Greater Himalaya, the Lesser Himalaya and the Outer Himalaya or Siwalik Range. Three kinds of antecedent streams are identified in terms of their antecedence to the aforesaid three ranges of the Himalayas. (1) The Indus, the Sutlej, the Ganga, the Ghaghra, the Kosi, the Brahmaputra etc., are antecedent to all of the three ranges of the Himalayas as they cut across the Greater, the Lesser and the Outer Himalayan ranges. It is believed that these rivers were present before the folding and upliftment of the Himalayan ranges. With the upliftment of these ranges during Tertiary period these rivers maintained their southward courses through downcutting their valleys in harmony with the rate of upliftment. All of these rivers have formed deep and narrow gorges while cutting across the Himalayan ranges. The Indus river has dug out a 17,000-foot deep gorge across the Greater Himalayas near Gilgit. It may be pointed out that a group of geologists does not believe in the antecedence of the Indus river. According to them the

Indus is an example of consequent stream because consequent streams might have developed on the southern slopes of the Himalayas. These rivers might have extended their courses northward through the process of headward erosion resulting into gradual backward shifting of their water divides. Ultimately, the streams of the southern slopes of the Himalayas might have captured the streams of the northern slopes and thus they might have extended their courses across the Himalayan ranges. The Arun river has been quoted as the best example of antecedent river by L.R. Wager (1937). The Arun river flows from west to east in its upper reaches following the structural grain of the region to the north of the Himalayas. It suddenly turns due southward and drains across the Himalayas through several deeply entrenched narrow gorges. A few geologists opine that the sudden southward bend of the Arun river is due to river capture. The advocates of the antecedent origin of the Arun river maintain that the convex course of the river across the Himalayas could be possible only through active downcutting of the valley in accordance with the rate of upliftment. A series of terraces on either side of the valley of the Arun river also validate its antecedent origin. (2) The rivers, which are antecedent to only two ranges of the Himalayas viz. the Lesser and the Outer (Siwalik Range) Himalayas, are the Jhelum, the Yamuna, the Ravi, the Ramganga etc. (3) The streams, which are antecedent only to the Outer Himalaya or Siwalik Range, include the Kamla, the Rapti, the Gola etc.

Superimposed Drainage System : Like antecedent streams superimposed streams are also not adjusted to regional geological structures and slope and thus are insequent or anti-consequent streams. Superimposed stream means a river which, flowing on a definite geological formation and structure, has inherited the characteristics of its previous form developed on upper geological formation of entirely different structural characteristics. In other words, a superimposed drainage is formed when the nature and characteristics of the valleys and flow direction of a consequent stream developed on the upper geological formation and structure are superimposed on the lower geological formation of entirely different characteristics. It is not necessary that the upper geological formation is conformal to the lower geological formation, rather it happens that some times the upper rock cover is entirely different from the underlying geological formation viz., the upper cover may consist of horizontally bedded sedimentary rocks while the lower cover may be composed of folded sedimentary beds, or domed structure or batholithic intrusives. The consequent streams are developed on almost flat ground surface of the horizontally bedded sedimentary rocks. These streams develop their valleys through vertical erosion (downcutting). With the passage of time the lower structure is exposed to the river which continues its downcutting and extends its valley downward on the lower geological structure (say anticlinal folds, domes or batholiths) and thus the valley developed on the upper structure is superimposed on the lower structure. The lower structure has to accept the form of the valley already developed on the entirely different upper structure. Thus, the river maintains the form of its valley, the flow direction and its drainage patterns as usual. Such rivers are called superimposed rivers. Had the lower structure been present on the ground, the nature of drainage system would have certainly been entirely different but the buried different geological structure has no alternative other than to adopt the nature of valley and flow direction developed on the upper structure. Such streams are unconformal to the local geological structure and slopes. Such unconformal valleys have been named as 'superimposed' by J. W. Powell (1857) and D. Maw (1886) and 'superposed' by W.J. McGee (1888).

The fundamental difference between superimposed and antecedent drainage systems lies in the fact that the former represents the superimposition of the valley developed in the upper geological formation and structure on the lower structure irrespective of its types and complexities but without any upliftment of the land area while the latter represents the maintenance of former course of the river and its valley inspite of upliftment of the land area across the river course due to concomitant valley deepening and upwarping of the land.

Most of the rivers of the Deccan trap (lava) region of Peninsular India are superimposed because new drainage system was evolved on the new surface formed due to cooling and solidification of lavas erupted during the Cretaceous and early Tertiary periods and these rivers were superimposed on the lower formation after the removal of lava covers. For example, the Subarnarekha river is superimposed on Dalma and phyllite hills to the west of Chandil in the southeastern Chhotanagpur plateau region of Bihar.

Example : The Son river flowing across the Khainjua ridges in the southern part of Rewa plateau (Madhya Pradesh) is a fine example of superimposed river. The Son river flows parallel and very close to the Khainjua ridge between 81°15' E longitude and Deoland. In fact, the Khainjua ridge of quartzite sandstones makes the southern bank of the Son valley. The river suddenly cuts across the Khainjua ridge at Deoland and then flows in south-easterly direction. After its confluence with the Banas river it again turns northward and cuts across the Khainjua ridge and flows in easterly direction.

The Son cannot be an antecedent river, though it cuts across Khainjua ridges at two places, because (i) the Khainjua ridges are composed of lower Vindhyanas which were folded and warped before the deposition of the upper Vindhyan sedimentaries, (ii) there are no evidences to support any localized rise in the ridges after Cambrian period, (iii) the evidence of laterites to the north of Kaimur scarps denotes the spread of basaltic lava of the Deccan trappean age of Cretaceous period which might have buried and obliterated previous draiange lines, and (iv) the absence of any significant stream to the north of the Son does not support idea of the Son being older than the Crectaceous lava flow.

Thus, the Son appears to be superimposed river. It may be argued that the Deccan lava flows during Cretaceous period covered the whole of the Rewa plateau and consequently buried the pre-existing drainage network and reliefs including Kaimurs and Khainjua ridges. With the initiation of sub-aerial conditions during Tertiary period new drainage lines began to develop over the solidified lava cover (basaltic lava) and the Son developed its present course through several stages. The lava cover might have been removed through continuous but gradual fluvial erosion. Once the Son established its course over the lava covered surface, it continued to erode its valley over the buried structures and thus its valley became superimposed over the Khainjua ridges.

Patterns of Drainage

The drainage pattern means the 'form' (geometrical forms) of the drainage systems and the spatial arrangement of streams in a particular locality or region. The location, number and flow directions of different streams of a particular region depend on the nature of slope, structural control, lithological characteristics, tectonic factors, climatic conditions, vegetal characteristics etc. Since there are much variations in the environmental conditions of different regions and hence there are also spatial variations in draiange patterns. Though the drainage patterns of some regions may be similar but net the same; but there are some common characteristics which enable us to distinguish different drainage patterns. Generally, the drainage patterns are divided into the following types: (1) trellised pattern, (2) dendritic pattern, (3) rectangular pattern, (4) parallel pattern, (5) centrifugal or radial pattern, (6) centripetal pattern, (7) annular pattern, (8) barbed pattern, (9) indeterminate or confused pattern, (10) herringbone pattern, (11) pinnate pattern etc.

Trellised Pattern : Trellised drainage patterns are formed by the network of tributaries and master consequent streams which follow the regional slope and are well adjusted to the geological structures. Such patterns are developed in the area of simple folds characterized by parallel anticlinal ridges alternated by parallel synclinal valleys. Several master consequent longitudinal streams are developed in the elongated longitudinal synclinal valleys. These longitudinal rivers and their valleys occupying the furrows between parallel ridges are parallel to each other. Several streams develop on both the flanks of the ridges and join the longitudinal synclinal streams at right angle. These tributary streams are called as lateral consequent streams. Thus, the resultant network of numerous longitudinal streams and transverse or lateral consequent is called trellised pattern. This pattern also resembles the rectangular pattern. In fact, the trellised and rectangular patterns are differentiated on the basis of spacing between the streams. If the streams are closely spaced, the resultant pattern becomes trellised while rectangular pattern is formed when the streams are widely spaced. Trellised drainage patterns are generally formed in the regions of cuesta topography where hard beds become escarpments and soft beds are eroded to become valley or lowlands.

Dendritic Pattern : Dendritic or tree-shaped drainage pattern is the most common and widespread pattern to be found on the earth's surface. The pattern is called dendritic on the ground that the network of tributaries of various orders and magnitudes of the trunk or master stream resembles the branches and roots and rootlets of a tree. The dendritic pattern is associated with the areas of homogeneous lithologies, horizontal or very gently dipping strata, flat and rolling extensive topographic surface having extremely low reliefs. Though dendritic pattern is independent of structural and lithological controls but almost uniform lithology (presence of same type of rocks in extensive area) presents most ideal condition for the development of dendritic drainage pattern. This pattern develops in a variety of structural and lithological environments such as in the mountainous and hilly areas (e.g. dendritic pattern is one of the dominant patterns in the Himalayas), on extensive plateau surface (e.g. Deccan plateau), on peneplain surfaces (e.g. the Peninsular peneplains of India, mostly in the basins of the Mahanadi, the Godavari, the Krishna, the Cauvery, the Damodar etc.), in the alluviated plains (e.g. Great Plains of North India), in the desert plains (e.g. Rajasthan desert), in the glaciated leoland regions (e.g. North European plains and northern lowlands of North America) etc.

The evolution of dendritic pattern is guided by the lithological characteristics mainly the permeability of underlying rocks, the amount and regime of rainfall and resultant surface runoff and the time factor. First of all the master or trunk consequent stream develops on gently sloping topographic surface. The master consequent gradually but slowly lengthens its longitudinal course through headward erosion. Major tributary streams of the master stream develop on either side of the trunk stream. These tributaries also lengthen their longitudinal courses through headword erosion and branches of tributaries are developed in due course of time. In the beginning the dendritic pattern is 'open' because the streams are few in number and are widely spaced but with the passage of time a dense mesh of multitudes of tributaries of various hierarchical orders is developed and thus the initial open network becomes a 'closed pattern' as the streams are very closely spaced. At a much later date, i.e., in the penultimate stage of drainage development, the dendritic pattern becomes 'simplified' due to 'surface abstraction' because many of the tributaries are integrated due to active river capture.

Slope and permeability of rocks very effectively control the number and extension of streams of dendritic drainage pattern. The drainage network of dendritic pattern becomes most extensive if the land surface is characterized by flat surface, level to gentle slopes and impermeable rocks. In such environmental conditions the pattern extends both in length and width but if the region is characterized by higher slope angles, the pattern extends more in length than in width. The extensive plateau surfaces of the Panna plateau, Bhander plateau and Rewa plateau (M.P.) composed of sandstones capping and the peneplained surface of the central Ranchi plateau composed of granite - genesis have favoured extensive dendritic drainage pattern while elongated dendritic pattern has developed along the middle and lower segments of the escarpments associated with these plateaus.

Rectangular Pattern : Rectangular pattern shows some degree of resemblance to trellised drainage pattern as tributary streams in both the patterns join their master streams almost at right angle but the confluence angle in the trellised pattern is guided by the strikes and dip angle of the rock beds while it is determined by the lines of weaknesses (e.g. faults and fractures and joints) of the rocks in the rectangular pattern. The second line of difference between these two drainage patterns is related to the spacing of streams i.e., the rectangular pattern is characterized by widely spaced tributaries while trellised pattern has a dense mesh of closely spaced tributaries.

Rectangular pattern is generally developed in the regions where the rock joints form rectangular pattern. The rocks are weathered and eroded along the interfaces of joints, fractures and faults and thus surface runoff collects in such long and narrow clefts (resulting from the weathering and erosion of joints) and forms numerous small rills. These rills are further extended in length and width and become channels. With the march of time a network of streams is developed wherein streams follow the lines of weakness (joints and fractures). The tributaries join their master streams almost at right angles and thus a rectangular drainage pattern is formed.

Radial Drainage Pattern : Radial drainage pattern, also known as centrifugal pattern is formed by the streams which diverge from a central higher point in all directions.

It is obvious that dome structures, volcanic cones, batholiths and laccoliths, residual hills, small table-lands, mesas and buttes, and isolated uplands favour the development of ideal radial pattern. The streams emerge at the central point of the aforesaid reliefs and drain down the slopes in all directions. Since the streams follow the slopes and hence they are basically consequent streams. These streams resemble the spokes of a wheel or the radii of a circle. If we take the entire drainage network of Sri Lanka, it exhibits the best example of radial drainage pattern at macro-level. The local upland situated to the south-west of Ranchi city has given birth to radial drainage pattern wherein the South Koel, the Subarnarekha, the Kanchi and the Karo rivers take their sources and radiate in different directions.

Hazaribagh plateau, Parasnath hill, Panchet hill and Dalma lava upland have issued ,radial drainage pattern. Morcha pahar and Lugu hill of Hazaribagh, Mt. Abu (Rajasthan) etc. present ideal examples of radial drainage pattern. At a regional level the Rohtas plateau of S.W. Bihar province presents a good example of radial drainage pattern wherein the Karamnasa and the Durgawati rivers and their source tributaries emerge from the south-western corner of the plateau and flow in the western, north-western, northern, and north-eastern directions whereas the tributaries of the Son river flow in the southern direction.

Centripetal Drainage Pattern : Centripetal or inland drainage pattern is opposite to the radial drainage pattern because it is characterized by the streams which converge at a point which is generally a depression or a basin. This pattern is formed by a series of streams which after emerging from surrounding uplands converge in a central low land which may be a depression, or a basin or a crater lake. The Kathmandu valley of Nepal presents an ideal example of centripetal drainage pattern wherein the tributary streams of the Baghmati converge in the tectonically formed circular basin. The depression formed at the top of Raigarh Dome in the Lower Chambal Basin has given birth to centipetal drainage pattern.

Annular Drainage Pattern : Annularpattern, also known as 'circular pattern', is formed when the tributaries of the master consequent streams are developed in the form of a circle. Such pattern is developed over a mature and dissected dome mountain characterized by a series of alternate bands of hard and soft rock beds. The differential erosion of hard and soft rock beds results in the truncation of the beds which produces ringed belted structure wherein relatively resistant beds project outward whereas the weaker (soft) beds form circular clefts. The master consequent streams emerge at the top of the dome and rediate in all directions down the slope like radial drainage pattern whereas tributary streams develop in the clefts formed due to erosion of soft beds, assume arcuate shape and join the master consequent streams and thus annular drainage pattern is formed. At a much later date tributaries of circular subsequent streams, which join the radial consequent, are also developed and thus the drainage pattern becomes a special case of trellised pattern. Annular drainage pattern has developed over denuded domes in the Weald of England. The mature dissected Sonapet dome of Bihar presents an ideal example of annular drainage pattern.

Barbed Drainage Pattern : Barbed drainage pattern, a rare kind of drainage pattern, is formed when the tributaries flow in opposite direction to their master streams. The tributaries join their master streams in a hook-shaped bend. Such pattern is generally developed due to river capture.

Pinnate Drainage Pattern : Pinnate pattern is developed in a narrow valley flanked by steep ranges. The tributaries originating from the steep sides of parallel ridges join the longitudinal master consequent occupying the valley at acute angles. The drainage network of the upper Son and Narmada rivers denotes the example of pinnate drainage pattern. This pattern resembles the veins of a leaf.

Herringbone Pattern : Herringbone drainage pattern, also known as rib pattern (like the rib bones of human beings), is developed in mountainous areas where broad valleys are flanked by parallel ridges having steep hillslide slopes. The longitudinal consequent streams, as masher streams are developed in the longitudinal parallel valleys while tributaries, as lateral consequent, after originating from the hillslopes of the bordering parallel ridges join the longitudinal consequent almost at right angle. The courses of the tributaries are straightened because of slope factor and little distance between the ridges and the longitudinal consequent occupying the valleys and thus the tributaries are not allowed to adopt sinuous course and join the longitudinal consequent at acute angles. The term herringbone has been derived from the pattern of bones of herring fish (mainly spine bones). The upper Jhelum river in the Vale of Kashmir receives numerous tributaries from both the sides and thus forms herringbone drainage pattern. The rivers occupying east-west trenches in the Himalayas form herringbone pattern. The Tamar Kosi, a left bank tributary of the Kosi river, the upper Rapti (a tributary of the Ghaghra river), the Rapti (another one), the left bank tributary of the Gandak river etc., have formed such drainage pattern.

Parallel Drainage Pattern : Parallel drainage pattern comprises numerous rivers which are parallel to each other and follow the regional slope. This pattern is more frequently developed on uniformly sloping and dipping rock beds such as cuestas or newly emerged coastal plains. The western coastal plains of India represent several examples of parallel drainage patterns where the streams after taking their sources from the western flanks of the Western Ghats drain in straight courses towards west to empty into the Arabian Sea. Parallel drainage pattern has also developed on the Eastern Coastal Plains of India. It may be pointed out that 'a subparallel pattern is therefore essentially an 'initial drainage pattern'.

River Capture

Meaning and Concept : The diversion of the part of the course of a river by another river is called stream diversion or stream capture or stream piracy. The river which captures the course of another river is called the capturing or captor stream while the part of the stream which has been divested of its course and water is called captured stream. River capture is a natural process which is more active in the youthful stage of the valley development because the streams are actively engaged in headward erosion and valley lengthening but river capture also occurs during mature and senile stages of the valley development through the process of lateral erosion and meander intersection. The stronger and more powerful streams (in terms of channel gradient, stream velocity and discharge and kinetic energy) capture the upper courses of weak and sluggish streams.

There are four major evidences of river capture viz. (i) elbow of capture, (ii) cols or wind gaps, (iii) water gaps, and (iv) misfit or underfit streams and valleys. The elbow of capture denotes the point where the course of the captured stream has been diverted to the course of the captor stream. Generally, the elbow of capture denotes sharp turn in the course of a river almost at right angle. The water gap denotes the deep and narrow valley in the form of a gorge formed by the captor stream through headward erosion across the ridge.

Wind gap is the dry portion of the beheaded stream just below the elbow of capture. The wind gap is also called as col. The misfit or underfit stream is the lower course of the captured stream. It is called misfit because of the fact that the former valley of the captured stream becomes too large and wide for the beheaded stream because of substantial decrease in the volume of water due to diversion of its water to the captor stream.

Ideal Conditions

Though the river capture is a natural process, but it does not take place in all circumstances rather it requires certain necessary conditions. In act, the process of river capture depends on channel gradient, depth of river valley, volume of water, velocity and discharge, lithological characteristics and geological structures, stage of cycle of erosion or the stage of river development. A particular river of a locality having deeper valley, more volume of water, steeper channel gradient and hence higher velocity and kinetic energy and flowing over less resistant and softer rocks than the other river of that region resorts to more powerful headward erosion than the latter, and thus may usurp the water and upper reaches of the weaker river. *It* may be, *thus*, inferred *that* river capture occurs under the following conditions - (1) steep channel gradient, (2) relatively narrow valley so that water may not spread in the otherwise wide and flat valleys, (3) higher volume of water so that velocity and discharge may be sufficiently high, (4) soft rocks so that the river may resort to rapid rate of headward erosion, (5) deeper valley than the valleys of other neighbouring rivers, (6) low sediment load so that the river may resort to active erosion etc.

Different Types

It is apparent from the above discussion that the process of river capture is effected by erosion of different sorts viz. headward erosion (valley lengthening), vertical erosion (valley deepening or downward cutting), lateral erosion (valley widening), and intersection of meanders (lateral erosion). Headward and lateral erosion is the most powerful geomorphological process of river capture. It may also be pointed out that headward erosion is more effective in terms of river capture during juvenile stage (youthful stage) of river development while lateral erosion becomes more operative during mature stage.

The capture of the course of a particular river by the other river through the intersection of meanders occurs mostly during late maturity and senile stage (old stage). Thus, the forms of river capture may be grouped in 3 broad categories viz. (1) capture through headward erosion, (2) capture through lateral erosion, and (3) capture through the intersection of meanders.

River Capture Through Headward Erosion : Most of the river captures occur due to headward erosion. In the initial stage of their development most of the streams and their tributaries are engaged in active headward erosion resulting into continuous creeping or shifting of water divides and lengthening of their valley thalwegs. The nature and intensity of headward erosion of any stream largely depends on the potential energy (height of the divide) and the steepness of the side slope of the water divide. Generally, the side slopes of the divide are unequal. The streams originating from the steeper slopes of the divide having relatively softer rocks and more precipitation and relatively short channel lengths degrade their valleys through the process of valley deepening more powerfully and resort to headward erosion at more accelerated rate than those streams which originate from the other side of the divide having less steep slope, relatively resistant rocks and low precipitation. Consequently, the erosive power of the former becomes much more than the latter. The powerful stream pushes the water divide backward towards the side of gentle slope through active headward erosion. Prolonged headward erosion by more powerful stream flowing on the steeper hillside of the divide results in the coalescence of the sources of both the streams on opposite sides of the divides. Since the valley floor of the stream of the steeper side of the divide is lower than the valley floor of the stream of the gentler side of the divide and hence the former captures the headwaters of the latter.

River Capture Through Lateral Erosion : Lateral erosion and consequent valley widening become more active and significant during mature stage of river development than vertical erosion and valley deepening. The water divides between parallel streams developed on sedimentary rocks of the coastal plains are gradually narrowed down due to lateral erosion and valley widening. The streams having more volume and discharge of water and relatively steeper channel gradient than the other streams resort to more lateral erosion due to which smaller parallel tributaries are consumed by the larger parallel streams. This process of river capture is called stream abstraction or natural selection.

River Capture Through the Intersection of Meanders : The streams adopt highly sinuous and meandering courses during their late mature and old tages of development because of the development of level to gentle slopes (0° - 5°) over major part of the area concerned. The meanders of two closely spaced streams are gradually sharpened due to continuous lateral erosion and ultimately they intersect each other and thus relatively more powerful stream captures the waters of the other stream. The Belan river, a tributary of the Tons river (which is itself a tributary of the Ganga river), has captured the lower course of its tributary the Seoti river near Deoghat (about 80 km south of Allahabad city) through meander intersection and now has pushed its course through the course of the Seoti while its older course, now an example of a palaeochannel, has become quite narrow due to sedimentation and anthropogenic processes (cultivation). Due to this unique process of river capture the confluence of the Belan-Seoti rivers has been pushed about 6 km upstream.

The older (palaeo) valley of the Belan now has become misfit valley.

Numerous cases of river capture have occurred in the Himalayan region. In fact, the present drainage system of the Himalayas is, to greater extent, the result of progressive stream piracy. The headward erosion has been the most active process of river capture in the Himalayas and the Western Ghats. The water divide between the headwaters of the Savitri river (draining into the Arabian Sea) and the Krishna river (draining into the Bay of Bengal) at the margin of the Mahabaleshwar plateau (Maharashtra) is very narrow. The Krishna river may capture the headwaters of the Savitri in near future. The process of river capture through lateral erosion and intersection of meanders is more active in the coastal plains and the Great Northern Plains (India).

Morphometry of Drainage Basins

'Measurement of the shape, or geometry, of any natural form—be it plant, animal or relief features—is termed morphometry' (A.N. Strahler, 1969) but in geomorphology 'morphometry may be defined as the measurement and mathematical analysis of the configuration of the earth's surface and of the shape and dimensions of its landforms' (J.I. Clarke, 1970). In fact, morphometry incorporates quantitative study of the area, altitude, volume, slope, profiles of the land and drainage basin characteristics of the area concerned (Savindra Singh, 1972). It has two distinct branches viz. (i) relief morphometry and (ii) fluvial morphometry. Relief morphometry includes the analysis of terrain characteristics through hypsometric curves, clinographic curves, percentage hypsometric curves, area-height curves, altimetric frequency histograms and curves, superimposed, projected and composite profiles, which assist in dealing with different aspects of landform characteristics of a drainage basin or of any geomorphic unit. Fluvial morphometry includes the consideration of linear, areal and relief aspects of a fluvially originated drainage basin. The linear aspect deals with the hierarchical orders of streams, numbers and lengths of stream segments and various relationships among them and related morphometric laws e.g., laws of stream numbers and stream lengths. The areal aspect includes the analysis of basin perimeters, basin shape (both geometrical and topological), basin area and related morphometric laws viz. law of basin area and law of altometric growth; stream frequency, drainage density and drainage texture. The relief aspect incorporates, besides hypsometric, clinographic and altimetric analyses, the study of absolute and relative reliefs, relief ratios, average slope, dissection index etc.

Historical Backdrop

Morphometric methods, though simple, have been applied for the analysis of area-height relationships, determination of erosion surfaces, slopes, relative reliefs and terrain characteristics as a whole since the beginning of the 20th century. But the vigorous application of statistical methods for the analysis of drainage basin characteristics started after the publication of classical research paper of R.E. Horton in 1945. The history of the application of morphometric methods has passed through three stages :

(1) The early phase of morphometric study in the 19th century was confined to the study of terrain depicted on small scale maps (macro morphometry) wherein heights and volumes of landforms and hypsometric and clinographic (height-slope relationships) characteristics were analysed.

(2) The second phase was marked with some improvements over previous phase in the beginning of the 20th century, wherein large-scale topographical maps were used for the study of specific relief features such as erosion surfaces, slopes, valleys etc. During 1930's Krumbein introduced simplified techniques for the analysis of geological data.

(3) The third phase of micro-morphometry started with the emergence of 'morphometric system' of R.E. Horton (1945) wherein morphometry was concerned with intensive study of small morphological units, particularly drainage basins. Numerous advanced statistical and mathematical methods have been applied since then to analyse the drainage basin characteristics. Now, aerial photographs and satellite imageries have enriched morphometric techniques.

Significant contributions in different aspects of relief and basin morphometry have been made by a host of scientists from time to time e.g. De Martonee (1934), P.S. Javanovic (1940), C.P. Peguy (1942, 1947 and 1948), R. de Smet (1951 and 1954), A.N. Strahler (1950, 1952, 1954, 1958, and 1969), P. Birot (1955), H. Baulig (1939, 1956, 1957, 1959), R.E. Horton (1932, 1945), J. Hanson-Lowe (1935), J.I. Clarke and K. Orrell (1958), M.A. Melton (1957, 1958,1959), S.A. Schumm (1956,1963), J.C. Maxwell (1955, 1960), H.W. Anderson (1957), D.R. Coates (1958), A.J. Brosco (1959), M. Morisawa (1959), C.W. Carlston (1963, 1965), 10 m. Brush (1961), R.J. Chorley (1957,1962,1966,1969,1972), L.B. Leopold and W.B. Langbein (1962), K.L. Bowden and J.R. Wallis (1964, 1966), R.B. Boyce and W.A.V. Clark (1964), A.E. Scheidegger (1965), W.C. Auckermann (1966), A.A. Miller (1955), V.C. Miller (1953), L.E.

Milton (1966), C.A. M. King (1966), R.L. Shreve (1966), K.J. Gregory (1968), K.J. Gregory and D.E. Walling (1968), V.T. Chow (1969), G.H. Comer and R.C. Zimmerman (1969), G.H. Dury (1969), A. Feniran (1969), P. Hagget and R.J. Chorley (1969), J.F. Daniel (1970), M.G. Wolman (1967),V. Gardiner (1971, 1973, 1974, 1976, 1977, 1978, 1979, 1981, 1982), W.B. Kirchner (1975), D.M. Mark (1974), P.M. Mather (1976), P. Mather and J.C. Doornkamp (1970), R.M. McCoy (1970), P.C. Patton and V.R. Baker (1976), K.S. Richards (1979), D.E. Walling (1971), J.S. Smart (19.72), J.S. Smart and A.J. Surkan (1967), K.J. Gregory and D.E. Walling (1968, 1971, 1973), K.J. Gregory (1968, 1976) etc.

'The study of morphometric characteristics of terrain began with the classic paper of R.L. Singh (1967) entitled 'morphometric analysis of terrain' and the publication of a research paper by B. Ghose et. al (1967) on quantitative geomorphology of Luni basin.' ...... A comprehensive research work in the form of a Ph.D. Thesis on 'drainage basin characteristics of the Belan river' was produced by Renu Srivastava in 1976' (Savindra Singh and S.S. Ojha, 1996). Significant contributions in the field of terrain and basin morphometry have been made by Bimal Ghose, S. Pandey, Surendra Singh, Amalkar (from CAZRI, Jodhpur), Renu Srivastava, Savindra Singh, S.S. Ojha (from Allahabad), K.R. Dikshit, V.S. Kale and S.R. Jog (Pane), A. Kumar (Bhagalpur), R.K. Rai (Shillong), H.S. Sharma (Jaipur), J.S. Rawat and R.K. Pande (Almora), Nageshwar Prasad (Burdwan) etc.

Criticism of the Theory

It is important to note that morphometry has received mixed reactions from the geoscientists at every stage of its development. It has been appreciated by many and at the same time it has also met with severe criticisms by others: As early as 1894 Penck drew the attention of geomorphologists towards the difficulty in accurate measurements of complex landforms of polycyclic origin. Hetner (1921) pointed out the weakness of morphometry in presenting the genetic explanation of landforms. Even Clarke and Orrell (1958) after assessing the reliability of a number of morphometric methods, made reservations about its utility and significance (R. Srivastava, 1976). Though the quantitative geomorphology has made a commendable progress and it has tried to fill up the conceptual vacuum created by the decline in the popularity of the Davisian model of landscape development, yet an alternative in terms of a broad systematic approach is still to be searched out as the quantitative wave has met with strong "objections from the geomorphologists who believe that quantification of much geomorphic data is either unnecessary or impossible" (S. W. Wooldridge, 1958).

Such extremist views have been expressed because of too much application of mathematical equations in this field. In fact, mere application of quantitative techniques without knowing their suitability and desirability may not lead to fecundity and productivity of the subject but some times its misapplication may lead to sterility and triviality (Savindra Singh, 1977) as 'quantitative and statistical methods offer no substitute for original thought about the problems, but they provide a valuable method of guiding the thought and reasoning into profitable channels and in assessing its validity' (C.A.M. King, 1966).

Though morphometry is a very sensitive and delicate tool (Savindra Singh, 1976) and when it is overcrowded with unnecessary and excessive mathematical formulae and equations, the results become confused and so the intuitive power of the investigator is lost, but when the useful and easier techniques, which are within the reach of geographers, are applied, morphometry certainly renders good results.

Geomorphic Unit

The main task before geomorphologists is to use an ideal unit of the earth surface for the study of its landforms. The search for an ideal areal unit, within which the collection, processing, organization and interpretation of data of the geometry of landforms, particularly of erosional origin, can be accomplished, has been the main aim of the geomorphologists right from Fenneman (1914) to R.L. Shreve (1966). 'Physiographic regions' (N.M. Fenneman, 1914), 'physiographic atoms' and 'drainage basins' have been selected as ideal areal units to serve the purpose. But, of late, fluvially originated drainage basin has gained currency as a 'geomorphic unit' because of its topographic, hydraulic and hydrological unity, which laid the keystone of Horton's 'morphometric system' (1945), wherein he emphasized quantitative analysis of drainage basin characteristics and erosional landforms (Savindra Singh and R. Srivastava, 1974). The limited, convenient and usually clearly defined and unambiguous topographic unit available in a nested hierarchy of sizes on the basis of stream ordering and an open physical system in terms of inputs of precipitation and solar radiation (Lee, 1964) of the drainage basin convinced R.J. Chorley (1969) to accept it as a fundamental geomorphic unit for the morphometric study of drainage basin characteristics. Since then drainage basin is being frequently selected as an ideal areal unit for the analysis of forms and processes of a region delineated by the basin perimeter.

Drainage basin is taken to be an open system as it is receiving energy or input from the climate over the basin and is losing energy or output through the water and sediment lost by the basin, largely through the basin mouth (K.J. Gregory and D.E. Walling, 1973) and thus adjustments and relationships between forms and processes are maintained.

The drainage basin receives energy from two sources viz. (1) from the atmosphere in the form of precipitation, and (2) from within the earth wherein endogenetic forces causing upliftment, warping, folding, faulting, tilting etc., provide energy (potential) to the erosional processes. On the other hand, the drainage basin loses energy (1) through evaporation, transpiration (from the vegetation of the concerned basin) and (2) outflow of water and sediment through its mouth.

Study of Drainage Basin

Though the drainage basin has recently been acclaimed as the basic unit in geomorphological study of fluvially eroded landscapes (R.J. Chorley, 1969), the use of this unit is very long established (K.J. Gregory, 1976). The exact origin of the concept of drainage basin as to who realized first the significance of forms and processes of the drainage basin, is hidden in the past history of geomorphological thought but P. Perrault (1674) is considered to be perhaps the first investigator, who visualized the significance of drainage basin, wherein he attempted to measure rainfall for three-year period to estimate drainage area and runoff. He, after his investigation, postulated that 'only one-sixth of the rainfall over the Seine basin was necessary to sustain river flow for a year' (K.J. Gregory, 1976). Phillippe Busche, probably for the first time, outlined the concept of the general topographical unity of the drainage basin (C.T. Smith, 1969).

By 1802, John Playfair appreciated the significance of drainage basin, at least by implication, and by mid-nineteenth century several people such as T.J. Taylor (1851) stressed the principle that the discharge of a river would depend upon the area of drainage basin (K.J. Gregory, 1976). W.M. Davis (1899) fully visualized the significance of rivers and presented the concept of the span of drainage basin as is evident from his own writing — *Although the river and hill-side waste sheet do not resemble each other at first sight, they are only the extreme members of a continuous series, and when this generatlization is appreciated, one may fairly extend the 'river' all over its basin and upto its very divide. Ordinarily treated, the river is like the veins of a leaf; broadly viewed it is like the entire leaf' (Davis, 1899).*

Jean Bruhnes used drainage basin as a parameter of the major division of France on the basis of the basins of Garonne, Lorie, Seine and Rhone Saone, as he realized the importance of water as a link between earth and man's activities (Smith, 1969) on the argument that 'Water is the sovereign wealth of a state and its people. It is nourishment; it is a fertilizer; it is power; it is transport...... watercourses are certainly not the only geographical connections of the earth with man, but they do present some principle or possibility of a linkage' (Bruhnes,1920). Drainage basin, accepted as 'natural unit', was taken to be substitute of 'political unit' as the real unit for geographical study. 'Though the concept of watershed as natural boundary for the geographical unit came into conflict but the basic tenet of the drainage basin as a natural division for the regional study could be preserved' (R. Srivastava, 1976). C. B. Fawcett (1917) realized the significance of drainage basin and divided England into 'provinces' mostly on the basis of drainage basins. 'Indeed from the period of primary settlement and colonisation to the industrial revolution and revolution in transport, the drainage basin, and more particularly stream networks had been directly linked with human activity in a variety of ways' (Smith, C.T., 1969).

It was the year 1932 when the drainage basin attracted the attention of R.E. Horton, an American engineer, who first of all presented an elaborate account of drainage basin characteristics and in 1945 he acknowledged the drainage basin as a 'morphometric system' wherein he applied morphometric techniques vigorously for the interpretation of salient features of drainage network. During 1950's and 1960's the morphometric proposals made by Horton were developed in the studies of drainage basin morphometry, which revealed regularities of stream networks and subsequently led to the studies of variation in network characteristics and to investigation of statistical or topological properties of network (K.J. Gregory, 1976). In fact, after the classical work of Horton in 1945, drainage basins attracted the attention of a large number of geomorphologists, engineers and hydrologists, who accepted the drainage network and the basin as a dynamic rather than a static unit. Consequently, significant contributions in the field of drainage basin study came from V.C. Miller (1953), A.N. Strahler (1953), J.C. Maxwell (1955), S.A. Schumm (1956), H.D. Anderson (1956), R.J. Chorley and D.E.G. Maim (1957), D.R. Coates (1958), R.J. Chorley (1957), A.N. Strahler (1957), M.A. Melton (1958), A.J. Broscoe (1959), M.A. Melton (1959), M. Morisawa (1959), C.W. Carlston (1960), J.C. Maxwell (1960), 10 m. Brush (1961), R.J. Chorley (1962), R.J. Chorley and M.A. Morgan (1962), L.B. Leopold and W.B. Langbein (1962), C.W. Carlston (1963), S.A. Schumm (1963), K.L. Bowden and J.R. Wallis (1964), R.B. Boyce and W.A.V. Clark (1964), A.E. Scheidegger (1965), C.A.M. King (1966), L.E. Milton (1966), K.J. Gregory and D.E. Walling (1971, 1973), R.L. Shreve (1966), R.J. Chorley (1969), L.B. Leopold, M.G. Wolman and J.P. Miller (1969), A.D. Abrahams (1970), V.Gardiner (1971), K.J. Gregory (1973), R. Srivastava (1976), Savindara Singh (1978), S.S. Ojha (1981), D.P. Upadhyay (1981), H.S. Sharma (1979), S.C. Mukhopadhyay, R.K. Rai etc.

Hydrological Cycle

The ground surface which supplies rain and or meltwater to a particular stream and its tributaries which drain that area is called drainage basin which is demarcated by well defined perimeter on the basis of water divides. Total network of master stream and its tributary streams of a particular drainage basin is collectively called drainage network which includes all types of streams viz. permanent (perennial), seasonal, ephemeral etc. Not only this, rills and figure-tip water paths are also included in drainage network. Drainage basin hydrological cycle includes input of water through precipitation, transfer of water (vertical and horizontal movement of water), collection of water in different storages, transfer of water from one storage to another storage and output (exit of water) of water through evapotranspiration, surface runoff, through flow, interflow, base flow and channel runoff.

The hydrological cycle of a drainage basin unaffected by man (it may be stated that most of the drainage basins of the world have been manipulated by man) operates in the following manner.

The basic input of a drainage basin is rainfall which is intercepted first by vegetation (if the basin has vegetation cover) and reaches the ground surface through the leaves, branches and stems of trees as 'aerial streamlets' or stem flow. The canopy of the forest is called interception storage where rainwater is retained by the leaves. Some portion of water of interception storage is evaporated while the re*maining portion* reaches the ground surface as stem flow.

Wherever the ground surface of the drainage basin is devoid of vegetation cover, the rainfall reaches the ground surface directly as through fall. Some portion of rainwater is lost to the atmosphere through evapo-transpiration from interception storage (leaves). The rainwater available on the ground surface forms surface storage of which a sizeable portion moves downslope as surface runoff, some portion is evaporated, some portion remains on the ground surface while some portion infiltrates downward into the soils to form soil moisture storage of which some portion is lost through evaporation and plant transpiration, some portion appears as seepage and springs via through flow and interflow while some portion percolates downward to form groundwater storage. Some portion of water from groundwater storage reaches the channel through base flow, some portion moves upward as capillary rise to 'soil moisture storage' and some portion is routed further downward through deep transfer. The channel storage receives water from surface storage through surface runoff, from soil moisture storage through interflow and through flow and from groundwater storage through base flow. Thus, the initial input of precipitation finds exit through two paths of output viz. (1) through evapotranspiration from interception storage, surface storage and soil moisture storage, and (2) through channel runoff from channel storage. Man affects and modifies the internal processes of hydrological regime of the drainage basin in a variety of ways.

Basin Morphometry

The basin morphometry includes the analysis of the characteristics of linear, areal and relief aspects of fluvially originated drainage basins.

Linear Aspects of the Basin : Linear aspects of the basins are related to the channel patterns of the drainage network wherein the topological characteristics of the stream segments in terms of open links of the network system (streams) are analysed. The drainage network, which consists of all of the segments of streams of a particular river, is reduced to the level of graphs, where stream junctions act as points (nodes) and streams, which connect the points (junctions), become links or lines wherein the numbers of all segments are counted, their hierarchical orders are determined, the lengths of all stream segments are measured and their different interrelationships are studied. The nature of flow paths in terms of sinuousity is equally important in the study of linear aspects of the drainage basins. Thus, the linear aspect includes the discussion and analysis of stream order (m), stream number (Nm), bifurcation ratio (R_b), stream lengths (Lm), length ratio (R_L), length of overland flow (L_g), sinuosity indices etc.

Stream Ordering : Stream ordering refers to the determination of the hierarchical position of a stream within a drainage basin. A river basin consists of its several branches (segments) having different positions in the basin area and they have their own morphometric characteristics and, therefore, it becomes necessary to locate the relative position of a segment in the basin, so that the hierarchical organization of stream segments is visualized. Thus, 'stream order is defined as a measure of the position of a stream in the hierarchy of tributaries' (L.B. Leopold, M.G. Wolman and J.P. Miller, 1969). It was Gravelius who made first attempt in 1914 to determine the orders of stream network wherein he attempted to trace the streams from the outlet to the source like an explorer.

(i) *Gravelius' Scheme of Stream Ordering* : He first, identified the trunk stream by tracing it from its outlet to its source on the basis of greatest width, discharge, headward branching and junction angles and thus the trunk stream was assigned the position of 1st order. By applying the same procedure he designated order 2 to all the streams which joined the trunk stream of 1st order. Next, all of the streams joining 2nd order streams, were designated by order 3 and this procedure was continued till the most remote finger-tip tributaries were assigned the highest order. This scheme of stream ordering by Gravelius was not appreciated as the ordering was based on subjective decision of the investigator at each bifurcation and the hierarchical orders were not symmetrically related to the magnitude of a given segment or link (M.J. Woldenberg, 1967 ; P. Hagget and R.J. Chorley, 1969).

(ii) *Horton's Scheme of Stream Ordering* : R.E. Horton, an American engineer, presented his scheme of stream ordering (1932, 1945) which was opposite to Gravelius' scheme. According to Horton ordering of stream begins from the finger - tip tributaries, which do not have their own feeders, rather they are independent in terms of supply of water. Such finger-tip streams are designated as 1st order streams. Two streams of first order, when join together, form 2nd order stream just below their junction. Similarly, two streams of 2nd order meet to make the stream of 3rd order and this process continues till the trunk stream is given the highest order.

Simultaneously, 2nd order streams may have other 1st order streams and 3rd order streams may have additional 1st order and 2nd order streams and so on. Thus, according to Horton's scheme when two streams of same order meet, they form the next higher order and each stream can receive tributaries of lower orders than its own order. In other words, the stream order increases only when two streams of same order join together. If a stream or streams of lower order join a stream of higher order, the order of the receiving stream does not increase. After all of the streams of a drainage basin are classified in the first round, they are reclassified in order to determine the headward extension of the streams of different orders except the 1st order streams. For example, one of the two streams, which form 2nd order stream, is extended headward on the basis of length, linear extension, discharge and some times by choice (when both the streams are of the same length) and is renumbered as 2nd order stream (previously it was 1st order stream). Similarly, one of the two streams, which form a stream of 3rd order, is extended headward to water divide and is renumbered as 3rd order stream. This scheme of reclassification continues till the trunk stream of the highest order becomes the longest drainage line of the basin.

Horton's scheme of stream ordering is difficult, tedious and time-consuming as it involves classification and reclassification of streams several times. Some of the finger-tip tributaries are given orders higher than one while other finger-tip tributaries of the 1st order with the same magnitude are not upgraded.

(iii) *Strahler's Scheme of Stream Ordering* : A.N. Strahler (1952, 1953, 1957 and in Chow's 1964) modified the Horton's scheme of stream or dering by removing the problem of reclassification and renumbering of streams. According to him ,each finger-tip channel is designated as a segment of 1st order. At the junction of any two 1st order segments, a channel of 2nd order is produced and extends down to the point, where it joins another 2nd order segment whereupon a segment of 3rd order results and so forth' (A.N. Strahler, 1969). These streams may have additional stream segments of lower orders than their own order and thus these do not affect the classification. It may be mentioned that the hierarchical order increases only when two stream segments of equal order meet and form a junction. The order does not increase if a lower order stream segment meets a stream segment of higher order. Strahler's scheme is popularly known as 'stream segment method'.

Thus, the objections labelled against Hortonian scheme was overcome by Strahler's modifications. The advantage of this simple scheme is that it can be derived mathematically from the concepts of elementary combinational analysis (M.A. Melton, 1959) as 'it designates all unbranched segments as the same order, and it gives highest order to one segment rather than to the whole of the trunk stream'

(K.J. Gregory and D.E. Walling, 1973). Strahler's scheme is simple and easy for application and it maintains the ordinal character of stream ordering and produces the same maximum basin order as that of the Horton's scheme but it suffers from the limitation (this is also in the case of Horton's scheme) in that the stream segments of lower order, when meet the trunk stream or the segments of higher order than their own, do not have the role to increase the order of the latter but the addition of a single 1st order segment in the upper reaches of the basin can raise the order of the trunk stream.

(iv) *Shreve's Stream-Link Magnitude Method* : R.L. Shreve (1966, 1967) proposed his scheme of 'stream-link magnitude' based on the 'interval scale' of stream ordering wherein each exterior link or 1st order segment is given a magnitude (M) of 1 and 'each successive link a magnitude equal to the sum of all the lst order segments, which ultimately feed it' (P. Hagget and R.J. Chorley, 1969). Thus, combination (*) of links M1 and M2 gives a downstream link magnitude of $M_1 + M_2$.

$$M_1 * M_2 = M_1 + M_2$$

Thus, in Shreve's scheme each segment (link) has its own contribution in increasing the magnitude (i.e. stream order) of the segment which it meets. For example, if two 1st order (magnitude) stream links meet, 2nd magnitude (2nd order) is formed down the confluence. When two segments of magnitude 2 meet, the magnitude increases to 4 (2 + 2 = 4) down the confluence. If a segment (link) of magnitude 1 meets the segment of magnitude 4, next higher magnitude (i.e. 5) is formed down the confluence (i.e. 4 + 1 = 5 magnitude) and so on.

(v) *Scheidegger's Scheme of Stream Order* : A.E. Scheidegger (1965) presented his scheme of 'consistent law of stream ordering' based on 'ratio scale measures' and involving four postulates of algebra of combination of stream segments which is cumulative as well as associative.

(a) When two similar segments (G') are combined, the resulting segment has its order increased by an integer,

$$G' * G' = G' + 1$$

(b) A combination of two segments of lower order (G' – 1) with a given order should increase the order of the latter by one integer,

$$G'*(G'-1)*(G'-1)=G'+l$$

(c) To validate the distributive law, it must be postulated that the sequence in which these segments join is immaterial,

$$[G' * (G'-1)] * (G'-1) = G' * [(G'-1) * (G'-1)]$$

(d) To satisfy the commutative law, it does not matter whether a G′ segment joins a G″ one, or vice versa,

$$G' * G'' = G'' * G'$$

From these four postulates a general law for the stream order magnitude has been suggested. In practical terms of consistent ordering the finger-tip tributaries are given an index of the exponent to base 2 (i.e. $2^1 = 2$) and all links are given an index number 1 and indexing of all downstream links are completed by adding the index numbers combining upstream at the junction to a given number 1 and thus each link is given an order magnitude (G) as follows—

$$G = \log_2 2M$$

Where G is desired magnitude and M is the magnitude of exterior links considered as magnitude 1.

Thus, the ordering of streams is completed in two stages (i) indexing of links, and (ii) assignment of magnitude values wherein 2 becomes 1, 4 becomes 2, 6 becomes 2.59, 12 becomes 3.59, 18 becomes 4.17 etc.

Though Scheidegger's scheme yields more classes of segments than Strahler's scheme and gives more accurate reflection of segment flow magnitude but it has been seldom adopted by the geomorphologists as it involves complicated procedures at several stages and consumes much time and labour. It also does not help in the computation of bifurcation ratio.

Besides, M.J. Woldenberg (1967) used Scheidegger's index and its relationship with basin bifurcation ratio to present his new scheme of stream ordering wherein he used W to indicate stream magnitude. He proposed the following equation for the determination of magnitude.

$$W = \frac{\log M}{\log R_b} + 1$$

when $M = 1/2$

R_b = bifurcation ratio

J. Lewin (1970) has proposed alternative ordering method either upon stream junctions (nodes) or upon paths. Recently, K.J. Gregory and D.E. Walling (1973) have postulated that, 'A simple method of ordering' useful in relation to process, is to visualize each segment in the network to be of order 1 and to develop ordering additively on this basis.'

After a critical evaluation of different schemes of stream ordering, it becomes apparent that the scheme of A.N. Strahler is simple and easy for application and it has been extensively used by the geomorphologists for the morphometric analysis of drainage network as it, having a slight modification over Horton's scheme, is suited for the explanation of different laws of drainage network as postulated by R.E. Horton and verified by Strahler himself.

Bifurcation Ratio : Bifurcation ratio (R_b) which is related to the branching pattern of the drainage network, is defined as a ratio of the number of streams of a given order (Nm) to the number of streams of the next higher order (Nm + 1) and is expressed in terms of the following equation—

$$R_b = \frac{N\mu}{N\mu + 1}$$

where $N\mu$ = number of streams of a given order

$N\mu + 1$ = number of streams of the next higher order

Table : Bifurcation ratio

Stream Order (μ)	*Number of streams* ($N\mu$)	*Bifurcation ratio (Rb)*
1	110	3.9
2	28	4.0
3	7	2.3
4	3	3.0
5	1	—

Bifurcatio ratio, a dimensionless property of the drainage basin is supposed to be controlled by drainage density, stream entrance angles (junction angles), lithological characteristics, basin shapes, basin areas etc. The studies of bifurcation ratios in different regions of the world by different geomorphologists have revealed the following behavioural patterns of bifurcation ratios and these patterns have been examined by Savindra Singh, S.S. Ojha and S.P.Agnihotri on the basis of study of 125 drainage basins of 6 physiographic regions having different geological structure with associated lithologies, vegetal cover, rainfall, soils etc. *viz.* Palaniau upland, Ranchi plateau, S.E. Chhotanagpur, Rewa plateau (M.P.), Bhander plateau (M.P.) and Belan basin (U.P. and M.P.) (Singh et. al, 1984)—

(1) E.Giusti and W.J. Schneider (1965) have shown that bifurcation ratios within a given region tend to decrease with increasing order because 'as order increases the percentage of streams that coalesce into a higher order tributary also increases...... this increase is due to the diminishing amount of area available.' 'This hypothesis

does not hold good in totality as except two basins in none of 125 basins bifurcation ratios register trend of consistent decrease with increase of order but on an average decreasing trend of bifurcation ratios with increasing order besides some exceptions in one or two orders of each basin is substantiated' (Singh et. al, 1984).

(2) 'Basins of equal order but variable areas tend to have the smallest bifurcation ratio in the smallest area; the ratios increase with increasing areas upto a certain size beyond which the bifurcation ratios tend to become constant' (E. Giusti and W.J. Schneider, 1965). This hypothesis is nom substantiated as 'the basins of equal area but lower order show more fluctuations in the mean bifurcation ratios than the basins of equal area but higher order with variable areas' (Savindra Singh em. al, 1984).

(3) 'Mean bifurcatio ratios show stable trend in a region of uniform geological structure and lithologies but they show variable trends over varying geological structures'. ...Singh et. al (1984) have remarked that 'geological structure and associated lithologies do not cause significant variations in bifurcation ratios and this observation holds parity with the conclusion of Miller (1953)'.

(4) 'Mean bifurcation ratios vary from about 2.0 for flat or rolling basins to 3.0-4.0 for mountainous, hilly dissected basins' (R.E. Horton, 1945). 'This hypothesis does not hold good as none of the basins draining over flat and rolling surface of the studied regions (as referred to above) show mean bifurcation ratios of 2.0, rather the values range between 3.0 and 5.0 in the same way as they range in the basins developed over hilly and highly dissected regions but it becomes clear that mean bifurcation ratios of the basins of mountainous and dissected regions are slightly more than the mean bifurcation ratios of the basins of the flat and rolling surfaces' (Savindra Singh, et. al, 1984).

(5) 'Mean bifurcation ratios register very small variation from region to region irrespective of structural control' (Savindra Singh, et. al, 1984). 'Following A.N. Strahler it may be postulated that mean bifurcation ratios show small variation from one region to another and such variations may be ascribed to chance variations'.

'Thus, it may be said that there is very little variation in bifurcation ratios from one region to another, and any pronounced variation, if it is, may be due to chance occurrence either because of environmental factor or morphometric control of the basin itself' (Savindra Singh et. al, 1984).

Law of Stream Numbers : The law of stream numbers relates to the definite relationship between the orders of the basins and stream numbers. R.E. Horton's law of stream numbers states (1945) *'that the number of stream segments of successively lower orders in a* *given basin tend to form a geometric series beginning with the single segment of the highest order and increasing according to constant bifurcation ratio.'* For example, if the master stream is of 6th order and the bifurcation ratio is 4, then the number of stream segments from higher to lower orders (i.e. 6, 5, 4, 3, 2, 1) would be 1, 4, 16, 64, 256 and 1024 respectively. The law of stream numbers is expressed in the following form of negative exponential function model

$N\mu = R_b^{(k - \mu)}$

where $N\mu$ = number of stream segments of a given order

R_b = constant bifurcation ratio

μ = basin order

K = highest order of the basin

Example :

(i) Number of stream segments of 1st order

$= N_1 = 4^{6-1}$

$= N_1 = 4^5 = 1024$

(ii) Number of stream segments of 2nd order

$= N_2 = 4^{(6-2)}$

$= N_2 = 4^4 = 256$

Horton further worked out the following equation (using constant bifurcation ratio) to find out the total number of the stream segments of the whole drainage basin—

$$\Sigma\mu = \frac{R_b{}^k - 1}{R_b - 1}$$

where k = highest order of the basin

R_b = constant bifurcation ratio

If the basin is of 5th order and bifurcation ratio is 4—

$$\text{So, } \Sigma N_\mu = \frac{4^5 - 1}{4 - 1}$$

$$= \frac{1024 - 1}{3} = \frac{1023}{3} = 341$$

Table: Hypothetical Stream Number

Basin Order (μ)	*Stream Number* $N\mu$	*Constant bifurcation ratio* (R_b)
1	256	4.0
2	64	4.0
3	16	4.0
4	4	4.0
5 = k	1	4.0

$\Sigma N_\mu = 341$

The following regression equation of negative exponential function helps in ascertaining the law of stream number—

$$\log y = \log a - bx$$

where y = number of stream segments

x = stream order

a = constant

b = regression coefficient

The regression line drawn on the basis of above regression equation involving number of stream segments (gully - rill segments) and stream order (gully order) of Deoghat gully, Allahabad District), plotted on semi-log graph paper, almost validates the Horton's law of stream number as stated above because the coefficient of correlation is —0.993 and the percentage variance explained is 98.65.

Table : Morphometric Data of Deoghat Gullies (1991)

Gully basin order (μ)	*Number of gully segments* ($N\mu$)	*Mean length* (*m*)	*Cumulative mean length* (*m*)	*Basin area* (m^2)
1	84	19.24	19.24	290
2	23	35.04	54.28	1560
3	7	46.85	101.13	5,980
4	1	256.00	357.13	56,130

Length Ratio and Law of Stream Length : The proportion of increase of mean lengths of stream segments of two successive basin orders is defined as length ratio (R_L) and is calculated according to the following equation—

$$R_L = \frac{\bar{L}_\mu}{\bar{L}_{\mu-1}}$$

when $\bar{L}_\mu = \frac{\Sigma L_\mu}{N_\mu}$

where $\bar{L}_\mu$ is the mean length of all stream segments of a given order (μ)

ΣL_μ is the sum of lengths of all stream segments of a given order.

N_μ is the number of stream segments of a given order. The following relationships are generally noted between stream lengths and basin orders—

(i) Total stream length of given order is inversely related to stream order i.e., total stream length decreases from the lower order to successive higher orders.

(ii) There is positive relationship between mean stream length and basin order i.e., mean length increases with successive increasing orders.

(iii) Horton's law of stream length (1945) states *'that the cumulative mean lengths* of *stream segments of successive higher orders in crease in geometrical progression starting with the mean length of the 1st order segments with constant length ratio'* and the following positive exponential function model of stream length has been suggested—

$$\bar{L}_\mu = \bar{L}_1 R_L^{(\mu-1)}$$

where $\bar{L}_1$ is the mean length of the 1st order.

R_L is constant length ratio.

It may be mentioned that this is a theoretical equation which may not be applicable in its totality for the drainage basins of natural system as the constant length ratio is of remote possibility in nature.

The regression line drawn on the basis of following regression equation of positive exponential function model involving cumulative mean gully lengths and gully basin order of Deoghat gullies (Allahabad district), plotted on semi-log graph paper, again almost validates the Horton's law of stream lengths as the coefficient of correlation is 0.999 and percentage of variance explained is 99.82.

log y = log a + bx

where y = cumulativemean lengths of gully segments

a = constant

b = regression coefficient

x = gully basin order

Sinuosity Indices : The shape of the open link in terms of geometric structure of drainage line involves the calculation of deviation of observed path (OL from the expected path - almost a straight line (EL) of a river from the source to the mouth. In fact, no river, in practice, shows straight path in terms of open link as many causative factors force the drainage line (stream) to deviate from the straight path. These factors may be geological and hydrological controls, dip angles, slopes, absolute and relative reliefs etc. Thus, sinuosity of a stream denotes the degree of deviation of its actual path from expected theoretical straight path (course). The analysis of deviation of the course of drainage line from the straight path, say sinuosity, may help considerably in studying the effect of terrain characteristics on the river course and vice versa. Simultaneously, the degree of sinuosity may give a vivid picture of the stage of basin development as well as landform evolution. A few models have been developed for the calculation of sinuosity indices as follows—

(i) **S.A. Schumm's model** (1963)

channel sinuosity $= O_L/E_L$

where O_L = observed (actual) path of a stream

E_L = expected straight path of a stream

On the basis of above equation Schumm identified 5 categories of channel sinuosity e.g. straight course when channel sinuosity is 1.0, transitional course, regular course, irregular course and tortuous course (when channel sinuosity is more than 20). (ii) J. E. Muller's model (1968) - Muller proposed his model of sinuosity index in terms of hydraulic and topographic sinuosity, which requires the measurement of channel length (CL), valley length (VL) and the shortest distance between the source and the mouth of the river (Air L), and calculation of a few indices.

$CI = \frac{CL}{AL}$ where CI = channel index

$VI = \frac{VL}{AL}$ where VI = valley index

AL = air length

$SSI = \frac{CL}{VL}$ where SSI = standard

sinuosity index

HSI = % equivalent of $\frac{CI - VI}{CI - 1}$

$$TSI = \% \text{ equivalent of } \frac{VI-1}{CI-1}$$

where HSI = hydrological sinuosity index

TSI = topographic sinuosity index

The value of unity (1.0) of standard sinuosity index (SSI) is indicative of straight river course whereas index values between 1.0 and 1.5 put the river in sinuous shape and the value more than 1.5 represents a meandering course.

The hydrological and topographical sinuosity indices (HSI and TSI) can be used as significant morphometric tools in searching out the causative factors of the sinuosity and also in determining the stage of the basin development. During early stage of basin development, if other factors remain constant, the TSI (more than 60 per cent) dominates over HSI and after the removal of pronounced relief controls in the late mature and old stages, the hydrological sinuosity (more than 60%) scores over topographical sinuosity. Thus, sinuosity indices explain the hydrological and topographical characteristics of drainage basins.

Meander Properties : 'Though J.E. Muller (1968) explained the controls of sinuosity of the river in a practical term in his model but it suffers from the shortcoming of arbitrary definition of meandering course of a river as pointed out by J.C. Brice (1964) who has presented his models for the calculation and derivation of different components of the meandering course of the river. The high degree of symmetry of meander length, meander height and form ratios is the characteristics of meandering properties which are applied 'to a sinuous stream that has, from place to place along its course, one or a series of symmetrical arcs, the length of which is related to the width of the stream' (J.C. Brice, 1964) (Savindra Singh, and D.P. Upadhyay, 1982). J.C. Brice's model of computation of meander properties involves calculation of symmetry of meander length, symmetry of meander height, symmetry of form ratio and repetition of symmetry as follows:

(i) Symmetry of (arc) meander length

$$= 100 - \frac{100\,(\text{mean deviation of arc length})}{\text{mean arc length}}\%$$

(ii) Symmetry of (arc) meander height

$$= 100 - \frac{100\,(\text{mean deviation of arc height})}{\text{mean arc height}}\%$$

(iii) Symmetry of form ratio

$$= 100 - \frac{100\,(\text{mean deviation of form ratio})}{\text{mean form ratio}}\%$$

(iv) Repetition of symmetry = $\frac{Na}{\Sigma Na} \times 100$

where Na = % of number of arcs whose length does not deviate more than 25% from the average length of meanders.

ΣNa = total number of meanders

Generally, the rivers, which have these values (symmetry of meander length, meander height, form ratio and repetition of symmetry) above 60 per cent, are termed as meandering rivers but some times terrain and lithological characteristics of rocks restrict the rivers to develop meanders of expected size. 'Repetition of symmetry is of vital significance in deciding the meandering properties of the streams wherein the higher percentage of repetition of symmetry is closely related to the meandering course whereas low percentage is indicative of straight course' (Singh and Upadhyay, 1982).

Length of Overland Flow : The length of overland flow, considered as a dominant hydrologic and morphometric factor, is 'the mean horizontal length of flow-path from the divide to the stream in a first order basin and is a measure of stream spacing and degree of dissection and is approximately one half the reciprocal of the drainage density' (R.J. Chorley, 1969). Length of overland flow is one of the most important morphometric variables, which affects the hydrological and topographic development of the basins. It is generally related to the stage of basin development, but no definite parameter or limit or scale of the length of overland flow has been suggested so far and thus qualitatively it is generally observed that the early stage is marked with maximum length of overland flow and mature and old stages register marked reduction in Lg.

The author agrees with the corollary of D.R. Coates (1958) that 'other factors being constant, areas more advanced into maturity appear to contain smaller overland flow-lengths than youthful areas' because a drainage basin on an average develops maximum stream segments in its late youth and early mature stages and thus minimum (shorter) length of overland flow is found. It is important to note that cyclic concept of basin development should not be given undue importance as length of overland flow is also affected by other factors viz. rainfall intensity, infiltration rate, soils, vegetation covers etc.

Stream Junction Angles : Geometry of angular property of drainage network is related to the junction angle of two stream segments which is defined as 'the angle, *projected* to the horizontal, between average flow-directions, determined by the ends of the stream segments extending from the junction point to an upstream point at a distance equal to 0.2 times the average length of the second order streams' (J.K. Lubowe, 1964). Lubowe has remarked that 'the stream junction angle increases as the order of receiving stream increases.' According to S.A. Schumm (1956) 'where the valley side slope is steep in relation to the gradient of the master stream, a tributary tends to join at almost right angle; but where the gradients of master stream and valley side slope are almost the same, the tributary virtually parallels the main channel, joining it at small angle'.

Thus, the stream junction angles may be correlated with the stage of basin development because 'the entrance angles in mature basins are significantly less than those in youthful basins' (S.A. Schumm, 1956).

Areal Aspects

Basin area is very important morphometric attribute as it is related to the spatial distribution of a number of significant attributes such as drainage density, stream frequency, drainage texture, slopes, absolute and relative reliefs, dissection index etc., that is why H.W. Anderson (1957) has termed it as a 'devil's own variable because almost every watershed characteristic is correlated with area.' The drainage basin area is delineated on the basis of water divides and the areas of all stream segments of each order are measured with the help of digital planimeter. All of the ground surface, which directly feeds the 1st order segments, are included in the areas of 1st order basins. The area of second order stream segments includes the area of 1st order segments plus the areas of inter-basins, which are triangular patches of ground surface contributing directly to the 2nd order segments. The same principle works for all the increasing successive order segments (Singh and Srivastava, 1974). Thus, the basin area becomes automatically cumulative from the 1st order to the successive higher orders.

The are aspects of the drainage basin include the study of basin perimeter, geometry of closed links i.e. basin shape, law of basin area, law of allometric growth, stream frequency, drainage density, drainage texture etc.

Geometry of Basin Shape : The geometry of basin shape is of paramount significance as it helps in the description and comparison of different forms of the drainage basins and it is also related to the functioning of the units of the basins and its genesis. The ideal drainage basin is usually of pear shape but since it is dependent on the size (of the basin) and the length of the master stream of the basin and basin perimeter, which are themselves dependent on other variables such as absolute reliefs, slopes, geological structure and lithological characteristics etc., and hence a wide range of variation in basin shape is bound to happen. Thus, various methods have been suggested to calculate the shapes of the basins. On an average 3 sub-categories of basin shapes have been recognized viz. (i) circular, (ii) elongated, and (iii) indented. A compact shape may be non-elongated, non-indented, or slightly indented, whereas a non-compact shape may be elongated, non-elongated, non-indented or highly indented. Different popular methods of computation of basin shape are as follows:

(i) **Horton's form factor (F) (1932)**

$$F - \frac{A}{L^2}$$

where F = form factor indicating elongation of the basin shape

A = basin area

L = basin length

The value of 'F' varies from O (highly elongated shape) to the unity i.e. 1 (perfect circular shape). Thus, the higher the value of F, the more circular the shape of the basin and vice-versa.

(ii) **Stoddart's (1965) Elipticity Index**

$$E = \frac{\Pi L^2}{4A}$$

where E = elipticity index

Π = 3.14

A = basin area

L = basin length

The value of 'E' varies from 1 to 0. It is apparent from these two equations that 'E' is inversely proportional to 'F'

(iii) **V.C. Miller's Circularity Index (C) (1953)**

$$C = \frac{\text{area of the basin}}{\text{area of the circle with same perimeter as the basin}}$$

or

$$C = \frac{4\Pi A}{\rho^2}$$

where ρ = basin perimeter

The value of 'C' varies from O (a line) to 1 (a circle). The higher the value of 'C', the more circular shape of the basin and vice versa.

(iv) **S.A. Schumm's (1956) Elongation Ratio (R)**

diameter of the circle with

$$R = \frac{\text{diameter of the circle with same area as basin}}{\text{basin length}}$$

$$\text{Thus } R = \frac{\sqrt{\frac{A}{\Pi}}}{\frac{L}{2}} = \frac{2}{\sqrt{\Pi}}\sqrt{\frac{A}{L^2}} = \frac{2}{\sqrt{\Pi}}\sqrt{F}$$

or

$$\left(F = \frac{\Pi}{4}R^2\right)$$

Thus, Schumm's 'R' is proportional to the square root of Horton's 'F'. The value of 'R' varies from O (highly elongated shape) to unity i.e. 1.0 (circular shape). Thus, the higher the value of R, the more circular shape of the basin and vice versa.

(v) **Chorley, Maim and Pogorzelski's (1957) Lemniscate Method (K)**

(based on the comparison of the basin with Lemniscate curve)

$$K = \frac{L^2}{4A}$$

The higher the value of 'K', the more elongated shape and vice versa.

It is generally believed that if all the factors controlling drainage basins development remain constant, the basin shape becomes more and more circular with the advancement of stages of cycle of erosion (and hence stages of basin development) and the basin assumes pear shape in the penultimate (old) stage of its development. But in most of the cases the shapes of the drainage basins do not appear to be the result of stages of basin development as topography and structure play important role in determining basin shape. For example, if a drainage basin is jagged between two parallel hill ranges, its shape will remain highly elongated even in the old stage of cycle of erosion.

Significant Morphometric Variables

'Basin area, basin perimeter and channel length are significant morphometric variables which determine the shape, size and genetic aspect of drainage basins. Basin perimeter can be directly correlated with the square roots of basin area and increase or decrease in the former indicates increase or decrease in the latter. The coefficient of correlation between these two variables of 30 small drainage basins (of S.E. Chhotanagpur, Bihar) standing at 0.99 is very strong positive correlation which is significant at 1 per cent probability level. Similarly, coefficients of correlation between basin perimeters and channel lengths and between square roots of basin areas and channel lengths being 0.89 and 0.98 respectively are again significant at 1 per cent of confidence level. The significant relationships among these variables speak the story of genetic aspect of basin development.

The first order segments of the basins notch down the watersheds from which they take their sources through headward erosion. This process helps in the proportionate increase of basin area and the backwearing of watersheds (water divides) favours the increase of stream length and basin area alike. The backwearing of divides simultaneously increases the basin perimeter. This law works in the region of homogeneous lithology and geological structure, climate and plant cover but significant variation in relief characteristics upsets the smooth functioning of basin development and introduces departures from the above law' (Savindra Singh and D.P. Upadhyay, 1982).

Area Ratio **:** Area ratio denotes proportion of increase of mean basin areas between two successive orders and can be calculated by the following equation as suggested by A.N. Strahler (1969)—

$$Ra = \frac{\bar{A}_{\mu}}{\bar{A}_{\mu-1}}$$

where $\bar{A}_{\mu}$ is mean area of a given order of the basin.

When $\bar{A}_{\mu} = \frac{\Sigma A_{\mu}}{N_{\mu}}$

where Nμ = number of all segments of a given order

ΣAμ = total area of all stream segments of the same order

Since the area becomes cumulative with increasing orders and hence area ratios decrease with increasing orders within the basin.

Law of Basin Area : It may be stated that the basin area becomes cumulative from 1st order to successive higher orders and the trunk stream (master stream) of the highest order represents the total area of the whole basin. A.N. Strahler (1969) paraphrased Horton's law of stream length into law of basin area and postulated that, *'the mean basin areas of successive higher strewn orders tend to form geometric series beginning with mean area of the 1st order basin and increasing according to constant area ratio'* and suggested the following equation of the law of basin area—

$$\bar{A}_{\mu} = \bar{A}_1 RA^{(\mu-1)}$$

where $\bar{A}_1$ is the mean area of the 1st order and

Ra is the constant area ratio.

The plot of mean basin areas on a constant ratio scale (logarithmic scalp) on the vertical axis against stream order on an arithmetic scale on the horizontal axis involving the following regression equation produces a straight line of regression of positive exponential model.

$$\log y = \log a + bx$$

where y is the mean basin area

a is constant

x is basin order

b is regression coefficient

The above law almost hols good in the case of Deoghat gullies (Allahabad district) as the coefficient of correlation is 0.994 and percentage of variance explained is 98.96 per cent. But the fitted function shows considerable underestimation in the case of basin areas of 1st, 2nd and 4th order gully segments and an overestimation in the case of 3rd order basin areas. It may be mentioned that whenever the area ratio within different orders of a particular basin becomes more or less stable, the law of basin area comes nearer to truth but when area ratio registers large variation, the law of basin area loses its validity. Thus, the relation of geometric series may not be tenable in crude sense but the linear relationship between these two variables is no doubt validated.

Law of Allometric Growth : The biological law of allometric growth, which states proportionate increase in all parts of the organs of living organism with time, of organs of animals and plants (wherein relative growth of an organ becomes constant fraction of relative rate of growth of the whole body of the animal or plant) has been introduced by geomorphologists (M. Morisawa, 1957, M.J. Woldenberg, 1966, A.N. Strahler, 1969 etc.) in the fluvial system wherein by analogy the measurable properties of a drainage basin, such as stream lengths and basin areas, are compared with the organs of growing animal or plant and increase according to allometric growth law.

The successive increasing orders with the advancement of fluvial cycle indicate increasing time wherein appearance of new 1st order finger-tip stream segments with accelerated rate of erosion during youthful stage increases the order of the older segments. Under ideal conditions of uniform geological structure and climate, the rate of increase (growth) becomes orderly and follows the law of allometric growth. A.N. Strahler has suggested that if cumulative mean lengths and mean basin areas of different basin orders are plotted on logarithmic graph paper (both horizontal and vertical scales being constant ratio scale) a straight line of regression of positive power function is formed.

$$\log y = \log a + b \log x$$

where y = cumulative mean length

x = mean basin area

The high degree of coefficient of correlation of 0.999 and percentage of variance explained (99.8 per cent) between cumulative mean gully lengths and mean gully basin areas of Deoghat gullies (Allahabad district) validates the law of allometric growth.

Stream Frequency : Stream frequency or drainage frequency is the measure of number of streams per unit area (may be square meter, square kilometer and so on). For the computation of stream frequency (SF), the basin is conveniently divided into grid squares (more commonally one square mile/kilometer) depending on map scale and areal coverage of the basin and the number of streams in each grid is counted, tabulated and quantified. The data of stream frequency are classified into certain categories depending upon the nature of data. The spatial pattern of stream frequency is studied through isopleth or choropleth maps. The general categories of stream frequency are (i) very poor (SF_{vp}), (ii) poor (SF_p), (iii) moderate (SF_M), (iv) high (SF_H) and (v) very high (SF_{VH}).

Drainage Density : Drainage density refers to total stream lengths per unit area. R.E. Horton (1945) defined drainage density as a ratio of total length of all stream segments in a given drainage basin to the total area of that basin and thus it can be derived as follows-

$$Dd = \frac{L_k}{A_k}$$

where L_K = total lengths of all stream segments of a basin

A_k = total area of the basin

Horton's method (as given above) yields only a single value (of drainage density) for the entire basin and hence it cannot be applied for the study of spatial variations of drainage density within a given basin. The simplest way to calculate drainage density on a regional scale is to divide the basin into grid squares of one square mile or one square kilometer each and to measure the total stream lengths in each grid square and to group the derived data into drainage density categories viz. (i) very low (Dd_{VL}), (ii) low (Dd_L), (iii) moderate (Dd_m), (iv) high (Dd_H) and (v) very high (Dd_{VH}) drainage density categories and to prepare isopleths for the study of spatial distributional pattern.

It may be pointed out that the measurement of stream lengths in grid square (with the help of opisometer or threads) is very tedious, difficult and time consuming procedure and hence 'the medium of measurement of stream lengths per unit area to calculate drainage density has diverted the attention of interested workers in this field to explore alternative methods of rapid estimation of drainage density viz. 'the probabilistic line intersection method' of C. W. Carlston and W B. Langbein (1960), T. Wilgat's method (1966) based on the mean distances from the nearest water course, 'the photomechanical method' of Vince Gardiner (1974), J.J. Donahue's method (1972) using the number of points on a regular grid that touch a stream as an index of stream length, digitization of drainage lines, Savindra Singh's drainage texture method (1976, 1978, 1981 etc.).

The spatial variation in drainage density has been related to precipitation effectiveness (M.A. Melton, 1957), vegetation index (R.J. Chorley, 1957), permeability of terrain (C.W. Carlston, 1963), climatic character (where soil, relief, rock type or vegetal cover affect precipitation, C.A. Cotton, 1964), rainfall intensity (R.J. Chorley and M.A. Morgan, 1962 ; M.A. Melton, 1957), structure, particularly rock type, relative ease in infiltration of precipitation into ground surface and vegetal cover (Savindra Singh and Renu Srivastava, 1974) etc.

Drainage Texture : 'An important geomorphic concept is drainage texture by which we mean the relative spacing of drainage lines' (G.H. Smith, 1950). Horton (1945) defined drainage texture on the basis of stream frequency (number of streams per unit area). In fact, the term 'drainage texture' has been used loosely and no successful attempt has been made to search out a quantitative parameter for its calculation. According to Savindra Singh (1976, 1978) the term drainage texture must be used to indicate relative spacings of the streams in a unit area along a linear direction. Thus, he attempted to find out a new parameter in terms of drainage texture (Savindra Singh, 1976, 1978) to replace the derivation of drainage density. According to him, 'drainage texture refers to relative spacing of streams per unit length in grid squares (one mile X one mile or one kilometer X one kilometer). The derivation of drainage texture is quick and easier method, as it involves only the counting of stream crossings along the four edges of each grid and its two diagonals rather than measuring the lengths as is done in drainage density. The following equation is suggested to calculate drainage texture :

$$Dt = As = \frac{1}{(t+p)/2} \quad \ldots \text{ Savindra Singh, 1981}$$

where Dt = drainage texture

AS = average spacing between two streams

$$t = \frac{(t_1 + t_2)/2}{\sqrt{2}}$$

when t_1, t_2 = number of intersections between the stream network and grid square diagonals

and $p = \dfrac{P_1 + P_2 + P_3 + P_4}{4}$

where P_1 to P_4 = number of intersections between the stream network and grid square edges.

The data of drainage texture, so derived (on the basis of above equation) are classified into five Dt categories e.g. (i) very coarse drainage texture (above 0.8, Dt_{vc}), (ii) coarse drainage texture (0.80.—6, Dt_c), (iii) moderate drainage texture (0.6—0.4, Dt_m), (iv) fine drainage texture (0.4—0.2, Dt_F), and (v) very fine drainage texture (0.2—0.001, Dt_{VF}), values indicate stream spacing in mile or kilometer in grid squares. Isopleths are prepared on the basis of computed data of drainage texture of a given basin for the study of its (drainage texture) spatial patterns.

Savindra Singh, V. Gardiner and S.S. Ojha. (1986) using a 'principal component analysis' technique for the analysis of 'spatial variation of drain age density in Palamau upland' came to the conclusion that drainage density and drainage texture have almost identical loadings on the stream component and appear to yield virtually identical information concerning this aspect of basin form. Savindra Singh (1981) obtained very high correlation coefficient (above 0.9) between drainage density and drainage texture during the comparative study of these two variables in Palamau upland (29 drainage basins), Ranchi plateau (23 basins), S.E. Chhotanagpur region (30 basins), Bhander upland in M.P. (15 basins) and North Aravalli region (31 basins) and concluded that, derivation and use of drainage density may be replaced by drainage texture as the latter is much effectively related to other morphometric variables as drainage density and its (of drainage texture) derivation is quicker and easier than drainage density. So, the five categories of drainage texture (Dt) i.e. very coarse, coarse, moderate, fine and very fine should be taken as representatives of extremely low, low, moderate, high and very high Dd categories respectively.

Relief Aspects

The relief aspects of the drainage basins are related to the study of three dimensional features of the basins involving area, volume and altitude of vertical dimension of landforms wherein different morphometric methods are used to analyse terrain characteristics, which are the result of basin processes. Thus, this aspect includes the analysis of the relationships between area and altitude (hyposometric analysis), altitude and slope angle (clinographic analysis), average ground slope, relative reliefs, relief ratio, dissection index, profiles of terrains and the rivers etc.

Hypsometric Analysis : Hypsometry involves the méasurement and analysis of relationships between altitude and basin area to understand the degree of dissection and stage of cycle of erosion. Area-height curves, hypsometric curves and percentage hypsometric curves are generally used to show the relationships between altitudes and area of the basin. The basic data required for the study of area-height relationships are areas between successive contours and their respective heights. The area may be measured with the help of planimeter or may be estimated by intercept method. The height is obtained from the contour map.

Area-height curves indicate actual areas between two successive contours and hence horizontal axis represents area in terms of percentage of total area and the vertical axis shows height. Though area-height curve denotes relationship between altitudinal zones and corresponding areas but this does not represent actual profile of terrain and hence breaks in slope are completely concealed.

Hypsometric curve is generally used to show the proportion of area of the surface at various elevations above or below a datom (F.J. Morkhouse and H.R. Wilkinson, 1967) and thus the values of area are plotted as ratios of the total area of the basin against the corresponding heights of the contours and hence the area is represented by cumulative percentage. Hypsometric curve also does not reveal the actual profile of terrain.

Percentage hypsometric curve (suggested by A.N. Strahler, 1952) involving two ratios of relative height (h/H, where h denotes height between two successive contours and H indicates total height) plotted on the ordinate and relative area (a/A, where a denotes area between two successive contours and A indicates total basinal area) plotted on the abscissa in terms of percentage cumulative has proved fruitful in providing a basis for recognizing stages of the cycle of erosion in a drainage basin (Savidhra Singh and R. Srivastava, 1975). It is important to note that the type of hypsometric analysis of the basin advocated by A.N. Strahler is not primarily designed to show erosion levels but is more for comparative purposes between one basin and another to establish stages of denudation that can be applied to any basin (C.A.M. King, 1966).

Hypsometric and erosion integrals (HI and EI) may be calculated either by planimeter or mathematically. The hypsometric integral is the ratio of volume or percentage of total volume of the basin area below the curve and thus it reveals the volume of the area unconsumed by the dynamic wheels of erosion whereas the erosion integral (EI) is a proportionate area above the curve and thus it indicates the volume of area which has been eroded by erosional processes. The hypsometric integral has been accepted as an important morphometric indicator of the stage of basin development. A.N. Strahler (1952) related the hypsometric integral of above 0.6 (or 60%), 0.6—0.35 (or 60%—35%) and below 0.35 (or 35%) to the youthful, equilibrium and monadnock (old) stages of basin development respectively. But more commonally youthful, mature and old stages are represented by hypsometric integrals of more than 60%, 60%—30% and less than 30% respectively. It may be pointed out that 'the low value (of hypsometric integral, below 30%) is only maintained as long as a few monadnocks give relatively great difference in height between the highest and the lowest points, when the monadnocks are consumed, the integral returns to about 0.4 to 0.6' (40% to 60%) (C.A.M. King, 1966). It may be mentioned that hypsometric integral is very delicate morphometric variable and hence it should be used with utmost care and field checks, otherwise it may render misleading results.

Clinographic Analysis : Clinographic curves represent average slopes between successive contours and thus present panoramic view of the terrain. Unlike area-height and hypsometric curves, clinographic curves reveal the breaks in slope and sudden changes in the relief of the area and they also present the general trend of the surfaces. The construction of clinographic curves require data of slope angles between successive contours, contour lengths, heights and areas between successive contours, for which the following techniques have been suggested—

(i) Finsterwalder's (1980) clinographic curve;

Average slope angle between two successive

$$\text{contours} = \tan\phi = \frac{CI \times L}{A}$$

where CI = contour interval

L = total length of contour

A = total area between contours

Finsterwalder substituted his clinographic curve (1980), wherein lengths of contours are plotted against their respective heights by another curve known as hypsoclinographic curve which incorporates hypsometric data and contour lengths alike. Cumulative area is plotted along the horizontal axis and cumulative contour lengths are plotted along the vertical axis. Average slope is calculated according to following equation—

$$\text{Slope angle} = \tan\phi = \left[\frac{h + L_1}{A_1}\right]$$

where h = height of the controur

L_1 = length of lower contour

A_1 = inter-contour area

De Smet suggested modification in the above scheme and recommended the following procedure to calculate slope angle between contours—

$$\text{slope angle} = \tan\phi = \left[\frac{h \times (L_1 \times L_2)/2}{A}\right]$$

where L_1 and L_2 = lengths of two successive contours

(ii) Hanson-Lowe's (1935) Clinographic Curve

Hanson-Lowe's method of the calculation of average slope between two successive contours is based on the axiom that contours are concentric circles and the average slope between two successive contours may be derived by the following formula:

$$\text{slope angle} = \tan\phi = \frac{\text{CI (in feet)}}{\text{Ad (in feet)}}$$

where Ad = average distance between two successive contours and CI is contour interval.

when Ad = rl - ru

where rl is the radius of lower contour and ru is the radius of upper contour.

$$r - \sqrt{\frac{a}{\Pi}}$$

where a is the area above the contour. The slope angles so derived are plotted from above downward. When slope angles are very low, then these are exaggerated by multiplying by a constant.

(iii) Mean Slope Curve of A.N. Strahler (1952)

Average slope angle between two successive contours is

$$= \tan\phi = \frac{\text{CI (in feet)}}{\text{Aw(in feet)}}$$

where Aw = average width between two successive contours and CI is contour interval.

$$\text{Aw} = \frac{A}{(L_1 + L_2)/2}$$

where L_1 L_2 = lengths of two successive contours. Clinographic curve is drawn by plotting the slope angles against the contour height from above and the process proceeds downward.

(iv) Average Inter-Contour Width Curve of De Smet (1954)

De Smet suggested to construct Clinographic curve by plotting average width between two successive contours along the horizontal axis and contour height along the vertical axis. Average width between each pair of successive contours is plotted as horizontal straight line against the concerned contour. The outer edges of horizontal and parallel lines of average width are connected by straight lines to form the curve.

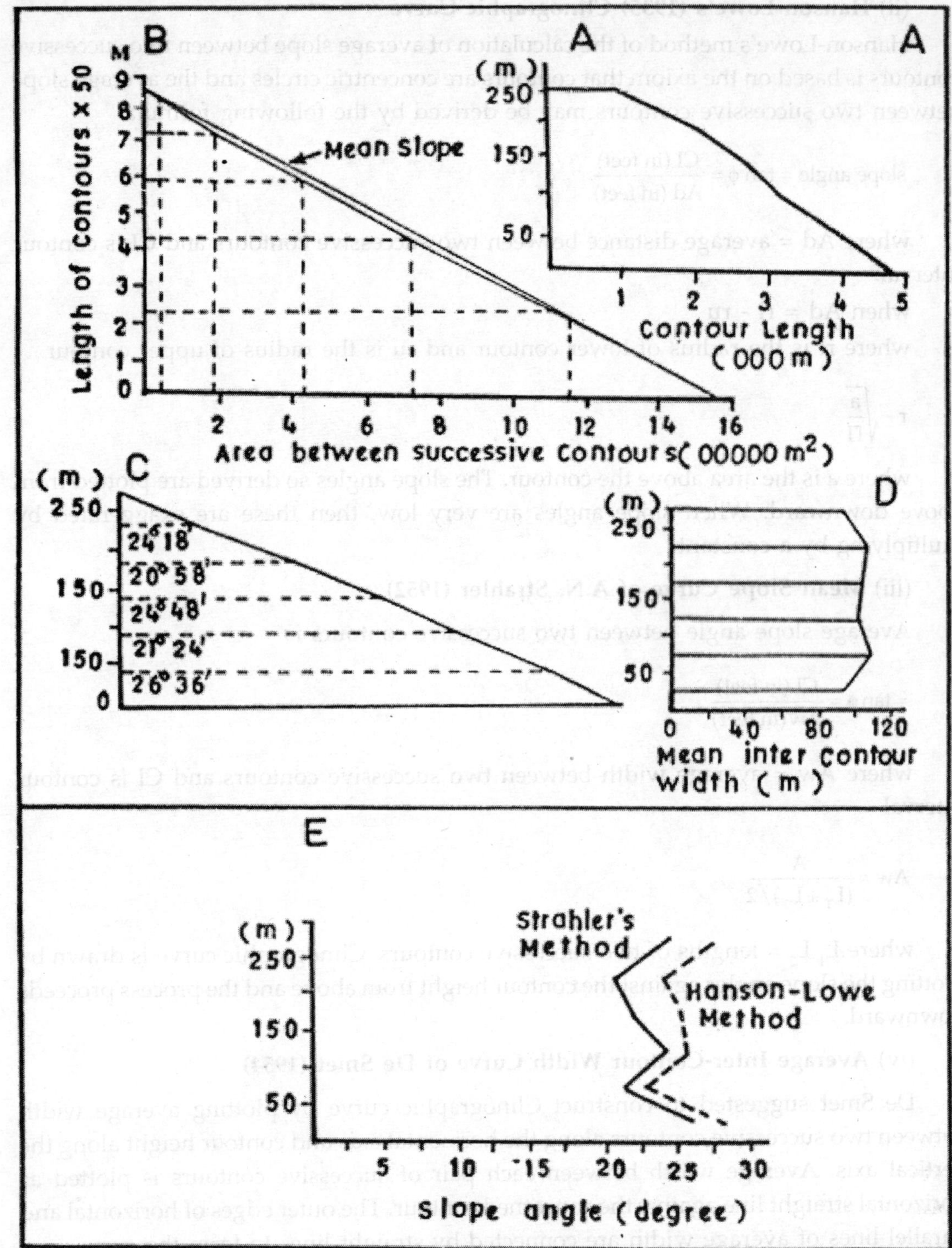
B
A
M
Length of Contours x 50
Mean Slope
Area between successive Contours(00000 m²)
(m)
250
150
50
1 2 3 4 5
Contour Length
(000m)
C
(m)
24° 18'
20° 58'
24° 48'
21° 24'
26° 36'
D
(m)
0 40 80 120
Mean inter Contour width (m)
E
Strahler's Method
Hanson-Lowe Method
5 10 15 20 25 30
Slope angle (degree)

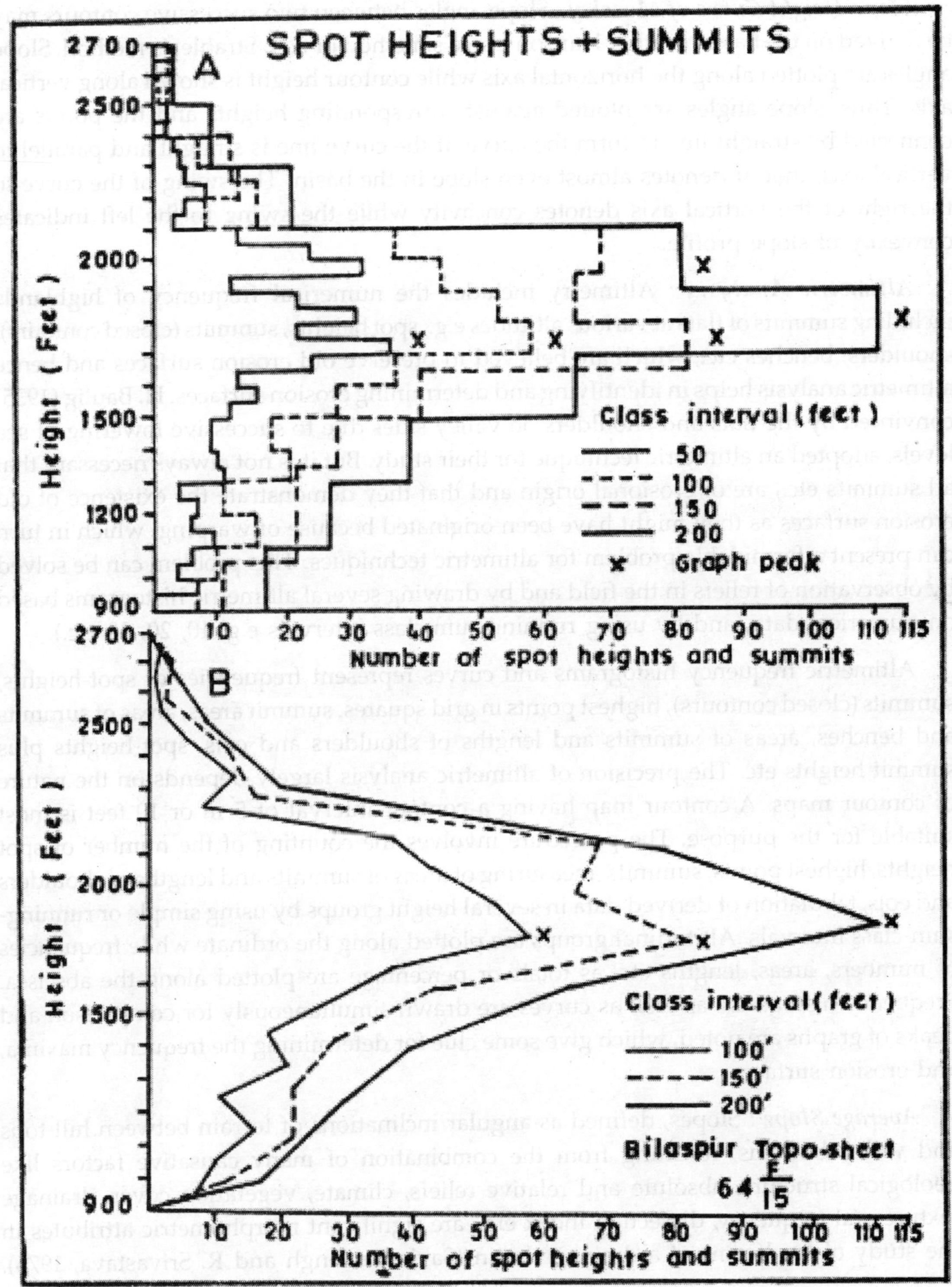
SPOT HEIGHTS + SUMMITS
A
2700
2500
2000
1500
1200
900
Height (Feet)
Class interval (feet)
50
100
150
200
x Graph peak
10 20 30 40 50 60 70 80 90 100 110 115
Number of spot heights and summits
B
2700
2500
2000
1500
900
Height (Feet)
Class interval (feet)
100'
150'
200'
Bilaspur Topo-sheet
64 E/15
10 20 30 40 50 60 70 80 90 100 110 115
Number of spot heights and summits

Slope-Height Curve of Moseley : Slope angles between two successive contours may be derived on the basis of either Hanson Lowe's method or AN. Strahler's method. Slope angles are plotted along the horizontal axis while contour height is shown along vertical axis. Thus, slope angles are plotted against corresponding heights and the points are connected by straight line to form the curve. If the curve line is straight and parallel to vertical axis, then it denotes almost even slope in the basin. The swing of the curve to the right of the vertical axis denotes concavity while the swing to the left indicates convexity of slope profile.

Altimetric Analysis : Altimetry includes the numerical frequency of highlands including summits of flats at various altitudes e.g., spot heights, summits (closed contours), shoulders, benches etc., which are believed to preserve old erosion surfaces and hence altimetric analysis helps in identifying and determining erosion surfaces. H. Baulig (1935) convinced by the flats and shoulders on valley sides due to successive lowering of sea levels, adopted an altimetric technique for their study. But it is not always necessary that all summits etc., are of erosional origin and that they demonstrate the existence of old erosion surfaces as they might have been originated because of warping, which in turn can present a formidable problem for altimetric techniques. This problem can be solved by observation of reliefs in the field and by drawing several altimetric histograms based on numerical data (and by using running-sum-class intervals e.g. 10, 20, 30 etc.).

Altimetric frequency histograms and curves represent frequencies of spot-heights, summits (closed contours), highest points in grid squares, summit areas, areas of summits and benches, areas of summits and lengths of shoulders and cols, spot-heights plus summit heights etc. The precision of altimetric analysis largely depends on the nature of contour maps. A contour map having a contour interval of 5 m or 10 feet is most suitable for the purpose. The procedure involves the counting of the number of spot heights, highest points, summits, measuring of areas of summits and lengths of shoulders and cols, tabulation of derived data in several height groups by using simple or running-sum-class intervals. Altitudinal groups are plotted along the ordinate while frequencies of numbers, areas, lengths etc. as totals or percentage are plotted along the abscissa. Frequency histograms as well as curves are drawn simultaneously for comparison and peaks of graphs are noted, which give some clue for determining the frequency maxima, and erosion surfaces.

Average Slope : Slopes, defined as angular inclinations of terrain between hill-tops and valley bottoms, resulting from the combination of many causative factors like geological structure, absolute and relative reliefs, climate, vegetation cover, drainage texture and frequency, dissection index etc., are significant morphometric attributes in the study of landforms of a drainage basin (Savindra Singh and R. Srivastava, 1975).

Computation of slope angles from topographical maps or through field measurement involves tedious and time-consuming procedures. Several techniques of the derivation and computation of average slopes from topographical maps have been suggested from time to time e.g. J.L. Rich (1916), C.K. Wentworth (1930), G.H. Smith (1938, 1939), E. Raize· and J. Henery (1937), W.C. Calef (1950), W.C. Calef and R. New Comb (1953), A.N. Strahler (1956), O.M. Miller and C.H. Summersoo (1960), R.J. Eyles (1965) etc., but the technique of Wentworth being easier and involving lesser measurement and calculation and more rapid procedure than other schemes has been widely used by several geomorphologists.

Wentworth has put forth his scheme of the calculation of average slopes in degrees wherein the contour map of the region is divided into grids of one mile X one mile (the grid may be one kilometer X one kilometer if the topographical map is on metric scale) each having straight lines at right angles and the numbers of contour crossings along the arms (edges) of grid are counted and then the average number of controur crossings per mile length (or per kilometer length) is computed. The same process is repeated by using oblique grids over the same contour map of the region and again the average numbers of controur crossings per mile (per kilometer) are counted and ultimately the mean of two calculations gives average number of contour crossings per mile (per kilometer) length. Using the following formula, as suggested by Wentworth, tan values are derived and actual slope angles resulting from tan values are obtained from log table.

$$\text{slope angle} = \tan\phi = \frac{N \times 1}{3361} \text{ for FPS system}$$

$$= \frac{N \times 1}{636.6} \text{ for MKS (metric system)}$$

where

N = number of contour crossings per mile or per kilometer length

I = contour interval

3361 = constant (for FPS system e.g. British scale, i.e. when the contour interval is in feet)

636.6 = constant for metric system (i.e. when contour interval is in meter).

The values of slope angles derived for each grid square of the drainage basin are tabulated and classified into convenient slope categories (e.g. (i) level slope category = 0°—2°, S_L ; (ii) gentle slope category = 2°—5°, SG ; (iii) moderate slope category = 5°—15°, Sms ; (v) steep slope category = 15°—30°, SS, and (vi) very steep slope category = above 30°, Svs) and isopleths are prepared for the study of spatial variations of average slope within the basin. It may be mentioned that Wentworth's scheme suffers from the drawback that cliff slope angles cannot be computed on the basis of his formula.

Relative Reliefs : Relative relief, also termed as 'amplitude of available relief' or 'local relief, is defined as the difference in height between the highest and the lowest points (height) in a unit area, it may be grid square, rectangle or a minute - grid square. Relative relief is a very important morphometric variable which is used for the overall assessment of morphological characteristics of terrain and degree of dissection. W.S. Glock (1932) used the term 'amplitude of relief' and defined it as 'the vertical distance from a horizontal fairly flat upland down to the initial grade of the streams'. M.A. Melton (1957) suggested to calculate relative relief by dividing the difference of height between the highest and lowest points in the basin (H) with basin perimeter (P, thus relative relief = H/P), whereas J.C. Maxwell (1960) defined relative relief as the quotient of maximum relief and basin perimeter (relative relief = maximum relief divided by basin perimeter) but these two schemes give only a single value of relative relief for the entire basin and hence isopleth map cannot be prepared with the result spatial variation of relative relief within the basin cannot be visualized. Therefore, grid method becomes more suitable and convenient for the purpose wherein the basin is covered with the mesh of grid squares (one grid being one kilometer X one kilometer) and relative relief in each grid square is calculated on the basis of the highest and lowest elevations and the data of relative relief so derived are tabulated and classified into six categories viz.

(i) extremely low relative relief = 0m — 15m, R_{EL} ;

(ii) moderately low relative relief = 15 m - 30 m, R_{ML} ;

(iii) low relative relief = 30 m - 60 m, R_L;

(iv) moderate relative relief = 60 m - 120 m, R_m;

(v) moderately high relative relief= 120 m- 240 m, R_{MH};

(vi) high relative relief= above 240m, R_H; and isopleth map is prepared for the study of spatial variation.

Dissection Index : Dissection index, expressing a ratio of the maximum relative relief to the maximum absolute relief, is an important morphometric indicator of the nature and magnitude of dissection of terrain. Slaucitajs (1936) used real area and projected area between successive contours to calculate dissection index between different height groups as follows—

$$DI = \frac{Ra - Pa}{Pa} \times 100$$

where Ra = real area between two successive contours

Pa = projected area between the same contours

De Smet (1951) suggested modification in the above scheme by including volume of relief and average slope as follows—

$$DI = \frac{Ra - Pa \times \text{average slope}}{\text{Volume}}$$

It may be mentioned that these two methods are time consuming and tedious procedures of the derivation of dissection index. Dov Nir (1957), there fore, presented a quick method of the calculation of dissection index which 'takes into account the dynamic potential state of the area—the ratio between relative altitude (relief energy) and the perpendicular distance from the erosion base'. He suggested the following formula for the derivation of dissection index—

$$DI = \frac{R_R}{A_R}$$

Where R_R is relative relief and A_R is absolute relief.

The grid method is used for the computation of dissection index. The values of dissection, so derived vary between 0 and 1 and the value of ratio can never be more than 1 except in the case of vertical cliff, and similarly O value is possible only in theoretical framework. Dissection index is also used as morphometric determinant of the stage of cycle of erosion wherein old, mature and young stages are related to dissection indices of less than 0.1 (less than 10 per cent), 0.1—0.3 (10 per cent — 30 per cent) and more than 0.3 (above 30 per cent) respectively. The values of dissection index derived for each grid of one square kilometer dimension are classified into 5 categories (e.g. (i) extremely low dissection index = 0—0.1, D_{EL} ; (ii) low dissection index = 0.1—0.2, D_L ; (iii) moderate dissection index = 0.1—0.3, D_M ; (iv) high dissection index = 0.3—0.4, D.; (v) very high dissection index = more than 0.4, D_{VH}) and isopleth map is prepared for the study of spatial variation.

Law of Channel Slope : The mean channel slope (gradient) is defined ,as the ratio of vertical drop to horizontal distance, measured from the upper end to the lower end of a single stream segment of a given order. The vertical leg of the triangle (every triangle denotes an order) is the average vertical drop ($\bar{H}_{\mu}$) of that order ; the horizontal leg is the average horizontal distance of that order and is identical with mean stream length, $\bar{L}_{\mu}$' (A.N. Strahler, (1969). Mean channel slope of each order is then derived and tabulated

Table : Mean channel slope (degree) of Deoghat gullies (Allahabad district)

Year	$\bar{S}_1$	$\bar{S}_2$	$\bar{S}_3$	$\bar{S}_4$	$\bar{S}_5$
1991	17.11	6.56	3.35	1.85	—
1992	17.85	6.90	2.70	1.47	1.75
1994	18.41	7.91	2.80	1.43	2.40

Source: Savindra Singh and A. Dubey, 1996, $\bar{S}_1 \ldots \bar{S}_5$ denote mean channel slope of 1st, 2nd, 3rd, 4th and 5th order gully basins.

'The law of channel slopes as propounded by R.E. Horton (1945) and applied by M. Morisawa (1959) and Leopold and Miller (1956) states that mean channel slope decreases with increasing successive orders in geometric series with constant slope ratio. The exponential function fitted to mean channel slope and stream (gully) order of the Deoghat gullies (Allahabad district) for the year 1991, reveals a perfect relationship as is evident from high coefficient of correlation of -0.99', (Savindra Singh and A. Dubey, 1996).

Year	*Slope ratio*	*% explained variance*	*'r' (degree of fit)*
1991	2.10	98.60	-0.99
1992	1.86	87.60	-0.93
1994	1.79	79.26	-0.89

Profile Analysis : Profiles, drawn from the contour map, are of great assistance to the geomorphologists who are primarily concerned with the analysis of relief and surfaces of terrain because profiles provide a visual perception of the actual nature of terrain but the profiles drawn from the contour map must be verified through field checks. Generally, profiles are basically of two types viz. (1) river profiles and (2) terrain profiles. River profiles are further divided into (i) longitudinal and (ii) transverse profiles while terrain profiles are classified into (i) serial profiles, (ii) superimposed profiles, (iii) projected profiles, and (iv) composite profiles.

Long profile or valley thalweg gives à vivid picture of breaks in longitudinal profile or longitudinal course of the river and numerous pronounced breaks may indicate nick points or heads of rejuvenation, and these breaks help in examining the polycyclic nature of landform development.

11

Regional Planning

Applied Geomorphology

Concept and Aspects : Each science subject has two main aspects of its study viz. (i) systematic and theoretical aspect, and (ii) applied aspect. The usefulness of any branch of knowledge lies in the fact that it helps in identifying and ameliorating the problems, faced by human society, arising out of interactions of human activities and geomorphic and environmental systems. From this stand point the study of the science of geomorphology or the science of landforms is of paramount significance as this science studies the genesis and development of surface features of the earth which provide the very base for all the biota in general and human being in particular. Most of the problems, which the present-day society is concerned with, are related to environmental degradation from local through regional to global levels caused by human activities related to rapacious exploitation and irrational utilization of natural resources consequent upon rapidly increasing population. Geomorphological knowledge may be more useful in understanding such environmental problems and paving way for their corrections and solutions.

According to D.K.C. Jones (1980) 'applied geomorphology may be difined as the application of geomorphic understanding to the analysis and solu tion of problems concerning land occupancy, resource exploitation, environmental management and planning.'

In fact, 'geomorphologists tend to consider all geomorphological research to be "applied." In the sense that each advance in knowledge provides a clear view of how the earth works, we must all be applied scientists' (R.G. Craig and J.L. Craft, 1982). 'The application of geomorphological techniques and the results of geomorphological investigations for resource evaluation, socio-economic development and mitigation of natural hazards and disasters etc. are significant aspects of applied geomorphology which should be taken into consideration by the geomorphologists' (Savindra Singh and S.S. Ojha, 1996). The main theme of applied geomorphology is to investigate the impact and changes brought in natural environmental systems and geomorphic/environmental processes by human economic activities, problems arising therefrom to suggest suitable remedial measures thereof; to study the impact of geomorphological processes / environmental processes on different aspects of human society and human activities ; and to attend the problems of resource management and monitor changes in the geomorphic systems / environmental systems including both landscape characteristics and landfom processes consequent upon human economic activities.

H. Th. Verstappen (1983) has outlined the following areas of interest where geomorphic knowledge and techniques may be applied for understanding and solving problems arising out of interactions of man with geomorphic / environmental system—

(1) Topographic and thematic mapping of natural resources.

(2) Survey of natural hazards, landslides, avalanches, earthquakes, vulcanism, land subsidence, flooding and droughts and natural hazard reduction.

(3) Rural development and planning emphasizing land utilization, control of man-induced soil erosion, channel manipulation and river basin (watershed) management.

(4) Urbanization, mining and construction.

(5) Engineering design.

Chorley's Classification

According to R.J. Chorley et. al (1985) 'application of geomorphology can be divided into two categories—

(1) Man as a geomorphic agent, in terms of his inadvertent and planned effects on geomorphic processes and forms.

(2) Geomorphology as an aid to resource evaluation, engineering construction and planning'.

   (1) The unintentional effects of human economic activities on geomorphic forms and processes include land use changes through cleaning of forests and burning of grasslands for crop cultivation; 'modification of terrain by engineering work, quarrying, mining, excavation and dumping; subsidence due to water extraction, oil drilling, mining and drainage' (Chorley et. al, 1985) ; accelerated rate of soil erosion consequent upon deforestation and

related increase in sediment load and recurrent floods and riparian decay of river banks; 'increased weathering rates due to pollution ; accentuated mass movements and slope failures due to engineering activities ; effects of hydraulic geometry due to changes of river load by dams or accelerated erosion ; interference with coastal debris movement; erosion and deposition by coastal engineering works; permafrost modification due to high latitude construction; and so on' (Chorley, et. al, 1985). On the other hand, planned effects of man's activities include those effects which arise out of several corrective measures to tame and modify geomorphic / environmental processes such as channel manipulation (e.g. construction of embankments, straightening of meandering courses of the rivers etc. to control recurrent floods); constructional measures to protect coastal areas against wave erosion (e.g. construction of seawalls, groins etc. to check cliff recession, to protect harbours from sedimentation by waves, to protect or to replenish sea beaches etc.) ;stabilization of sand dunes by plantation; channel diversion ; checking of soil erosion through afforestation and reforestation etc.

(2) 'The resource evaluation and planning aspects of applied geomorphology are concerned with such matters as the role of geomorphology in resource inventories and environmental management; soil and land evaluation, (D. Dent and A. Young, 1981); the production of maps for hydrological, erosional and stability control; geomorphic mapping, the mapping of land systems and terrain evaluation (C. Mitchell, 1973) concerned with the acquiring, classification and retrieval of terrain and other information for the use of civil and military engineers, earth scientists and planners (Chorley et. al, 1985) ; preparation of natural hazards maps on the basis of mapping ; urban planning in different geomorphic environment; construction and management of road network ; morpho-agricultural regionalization and land use planning etc.

Several branches of applied geomorphology have been developed during the last few decades viz. environmental geomorphology, urban geomorphology, anthropogeo morphology etc. which are directly related to the responses of interactions of man with certain sets of geomorphic / environmental systems and remedial measures for tackling the problems arising out of those responses.

'Urban geomorphology, a recent but more useful branch of applied geomorphology, is the study of landforms and their related processes, material hazards, in ways that are beneficial to planning development and management of urbanized areas, or areas where urban growth is expected' (R.U. Cook and J.C. Doornkamp, 1982). In fact, 'urban geomorphology deals with the lithological and topographical characteristics and geomorphic processes and hydrological conditions which determine the size and rate of urbanization and stability of urbanized localities, the impact of urban development on environment and geoenvironmental problems emanating therefrom' (Shalini Singh, 1996).

'Environmental geomorphology' deals with the impact of natural processes (e.g. vulcanism, floods and droughts, landslides, denudation processes etc.) on human activities and human landscapes on the one hand, and the manipulative impact of human activities on natural /environmental processes, problems arising out of these interactions, and remedial measures thereof leading to environmental management and natural disaster reduction, on the other hand.

Anthropogemorphology deals with anthropogenic activities as potent geomorphic process which can (and is) accelerate or decelerate the rate of natural geomorphic / environmental processes many fold which may result in severe geoenvironmental problems (and has already caused such problems) which become detrimental to human society. Almost all of the geomorphic processes/ environmental processes have been greatly affected and modified by human activities and a few gemorphic systems have become fragile e.g. occurrence of earthquakes, landslides, floods, droughts, stability of hillslopes and ground surface etc. have been greatly affected by man.

Indian Applied Geomorphology : 'It is heartening to note that several geomorphologists, mainly at CAZRI (Central Arid Zone Research Institute), Jodhpur, have focused their attention towards this aspect of geomorphology. The effects of processes and extent of erosion on land use in central Luni Basin (Rajasthan) were assessed by B. Ghose et. al as early as 1968.

The study of relationship between geomorphology and settlement and geomorphic controls on rural settlements is a significant area of applied geomorphology and such study was initiated by R.L. Singh at B.H.U., Varanasi during 1960s wherein relationships between landforms and rural settlements in different parts of the country were examined. Later on, the impacts of geomorphic influences on rural settlements were examined by P.C. Vats et. al (1976), P.C. Vats and Surendra Singh (1980, in Nagaur Tahsil of Western Rajasthan), Surendra Singh and P.C. Vats (1979, in Jodhpur district of Rajasthan), R.K. Rai (1982, in Meghalaya), Savindra Singh and O.P. Singh (1976, in Palamau upland, Bihar), Surendra Singh (1995, in arid environment of Rajasthan)' (Savindra Singh and S.S. Ojha, 1996).

'The influence, control and application of geomorphological characteristics on land use and agricultural planning (a significant theme of applied geomorphology) has been traced out by A.K. Sen and S. Singh (1977: significance of geomorphic factors on land use planning and development in Bikaner, Rajasthan), B.L. Bhar and M.N. Jha (1977 application of geomorphology to forestry), K.S. Bhatia and R.S. Singh (1976: evaluation of rainfall intensities and erosion index values for soil conservation), K.S. Bhatia and H.P. Chaudhary (1977 runoff and erosion losses and crop yields from slopy and eroded alluvial soils of U.P.), H.S. Sharma (1979 : physiography of lower Chambal valley and its agricultural development), Anita Kar and Amalkar (1981 :relevance of geomorphology in agricultural planning), R.K. Pandey (1981 : influence of geomorphology on land utilization in Gagas basin, Almora district, U.P.), S. Singh (1985 : role of geomorphology in the management of land and water resources in arid environment of Jodhpur district, Rajasthan; 1987: impact of geomorphology in land use planning in the Rajasthan desert), B.L. Sharma (1982: geomorphic control on agricultural taxonomy), P.C. Vats and S. Singh (1983: impact of geomorphology on land utilization in western Rajasthan), R. K. Pandey (1988: land use and slope instability in Lesser Himalaya), Savindra Singh and Neera Rastogi (1992 : morpho-agricultural regionalization, eastern Rewa Plateau, M.P.) and many more' (Savindra Singh and S.S. Ojha, 1996).

The influence of granite weathering and its differentiates on dam foundation in central India (D.K. Ghose, 1982) ; relevance of geomorphology in the exploration and planning of water resources in Indian desert ; significance of geomorphology on resource appraisal and regional and spatial planning (N. Prasad and A. Mahto, 1982) ; role of photo-geomorphological techniques in the integrated natural resources survey of the central Luni basin of Rajasthan (S. Singh, 1982) ; significance of geomorphology in spatial planning of an arid environment-Jodhpur district (S. Singh, 1985) ; geomorphology in the appraisal of natural resources (S. Singh, 1995) etc. (Savindra Singh and S.S. Ojha, 1996) are significant aspects of applied geomorphology which have been studied by Indian geomorphologists.

Environmental geomorphology of the trans-Yamuna region of Allahabad district, U. P. (A. Dubey, 1985), environmental management of coastal dunes (M.K. Bandopadhyay), environmental changes in the upper Chandra basin of Lahul Himalaya (M.K. Bandopadhyay, 1984), geomorphology and environmental management of Ravi Basin in N.W. Himachal Himalaya (D.K. Ghose, 1982), morphogenetic processes, landforms and environmental planning in Bhander plateau, M.P. (Savindra Singh and R.S. Pandey, 1983), flood hazards and environmental degradation in the Gomti Basin, U.P. (Savindra Singh, 1983), drainage system and flood hazards in Bharatpur (G.P. Gupta, 1982), sand movement and control of aeolian hazard (Amalkar and D.C. Joshi, 1995), evaluation of present-day geomorphic processes in the Indian desert for assessing land degradation and natural hazards (Amalkar, 1989), morphoecological variations in the Sundarban mudflats (M.K. Bandopadhyay et. al, 1987), wet coastal ecosystem and its conservation in west Bengal (M.K. Bandopadhyay, 1989), geoenvironmental problems in subhumid tropical environment (Savindra Singh and S.P. Agnihotri, 1989), landforms and micro-environmental management (D.K. Singh, 1988), gully erosion and management (Savindra

Singh and A. Dubey, 1996) etc. are significant areas of environmental geomorphology studied by Indian geomorphologists.

Thus, it is evident that the main areas where geomorphological knowledge and techniques have been applied by Indian geomorphologists include rural settlement and urbanization, land use and agricultural planning, soil erosion and conservation, management of water resources, terrain evaluation, land utilization, morpho-agricultural regionalization, resource appraisal, regional and spatial planning, environmental management, flood hazards and management, aeolian hazards, gully eorsion and management etc.

Regional planning and development is topmost priority of every developing country. 'Planning is considered to be a process of development having two main objectives e.g. (i) achieving an overall growth of the society and (ii) removing socio-economic disparities through the exploitation and utilization of all types of resources whether natural or human' (Savindra Singh, 1991). Planning requires suitable planning unit in terms of spatial unit which poses a great problem before the planners and decision makers. Generally, administrative units ranging from state (province) through districts to development blocks are selected for regional planning and development. This approach may be convenient from administrative point of view but this is not logically sound as any administrative unit cuts across the physiographic unit/natural unit and therefore the continuity of natural resources is broken which presents impediment in the exploitation and utilization of natural resources.

The attention of planners was drawn towards dynamic geomorphology in the USA in 1930s and hence drainage basin was chosen for overall man agement of the problems of the entire basin cutting across administrative units. Consequently, the Tennessee valley Authority (TVA) was commissioned in 1933 for total watershed management including control of soil erosion and flood, proper land use and land management, reforestation, eradication of deadly diseases (i.e. malaria, typhoid and tuberculosis) which were widespread in the basin in the beginning, navigation of the river, generation of hydroelectricity etc. After unprecedented success of the TVA the multi-purpose river projects were launched in the USA and other parts of the world e.g. the Missouri Valley Authority (MVA in the USA), the Damodar Valley Authority (DVA) in India etc. In fact, a drainage basin (watershed) represents an ideal geomorphic unit characterized by direct relationships between basin reliefs, fluvial process and man. The geoenvironmental and human problems are more or less uniform throughout the basin because of geomorphic and hydrologic uniformity. For example, several social problems and evils have taken root in the Chambal drainage basin (M.P. and U.P.) because of ravination and development of circuitous dense network of deep gullies and frequent severe floods. Most of the area of the basin has become badland due to ravination and hence is unsuitable for cultivation. Thus, 'the ravinous tracts of the Chambal valley have not only moulded the social behaviour of the local people but also of the adjoining areas as the circuitous network of ravines provides ideal hide outs and the people after committing crimes easily escape to these ravines which extend standing invitation to criminals and dacoits by providing safe shelter though with least comfort' (Savindra Singh, 1991). The knowledge of the mechanism of processes of gully erosion and ravination may be helpful for gully management and improvement and reclamation of ravinated area of the basin.

Even if the regional planning is attempted by selecting administrative units as planing units, the detailed information of topographic details, soils, natural resources etc. may be helpful for the purpose. Terrain evaluation and terrain classification, fluvial characteristics of the basin (channel flow, discharge, channel morphology, runoff etc.), groundwater condition etc. provide useful information to the decision makers and planners.

Hazard Management

Those events or accidents, whether caused by natural processes or human factors, are called extreme events which occur very rarely and agravate natural environmental processes to cause disaster for human society such as 'sudden tectonic movements leading to earthquake and volcanic eruption, continued dry conditions leading to prolonged droughts, floods, atmospheric disturbances etc. Environmental hazards may be defined as those extreme events either natural or man-induced, which exceed the tolerable magnitude, make adjustment difficult and cause immense loss to man such as volcanic eruptions, earthquakes, floods, landslides, accelerated erosion etc. Hazards are generally taken as processes, both natural and anthropogenic whereas disaster is a sudden adverse or unfortunate extreme event which causes great damage to human beings as well as plants and animals. Disaster occurs rapidly, instantaneously and indiscriminately. It, therefore, becomes obvious that the environmental hazards are processes whereas environmental disasters are the results or responses of environmental hazards' (Savindra Singh, 1991).

'A geomorphic hazard can be defined as any change, natural or man-induced, that may affect the geomorphic stability of a landform to the adversity of living things' (Chorley et. al, 1985). Geomorphic hazards range from very long-term hazards (e.g. crustal instability caused by earth movements leading to mountain building, faulting, folding, warping, subsidence, upliftment ; by changes in vegetation cover and type and hydrologic regime due to climatic change etc.) to short-term (e.g. volcanic eruption, earthquakes, floods, landslides, avalanches, riparian decay, changes in discharge, sediment load and runoff of a river consequent upon land use changes and management etc.).

Geomorphic knowledge helps in the identification, prediction, and assessment of probable effects and management of natural geomorphic hazards. A few characteristic features of volcanoes and their eruptions based on close observations of past volcanic eruptions are used as warning signs of impending eruptions. Regular measurement of seismic events and earth tremors by seismic methods ; regular measurement of ground surface mainly tilt measurement by tilt meters ; constant measurement of temperature of crater lakes, hot springs, geysers, fumaroles ; monitoring of gases coming out of craters, hot springs, geysers; monitoring of changes in the configuration of dormant or extinct volcanoes by lasers; measurement of local gravity and magnetic fields and their trends etc. help in making predictions of possible eruptions in the areas having past case histories of vulcanism. The path of lava flow known as lahar can be better predicted on the basis of detailed analysis of topography and identification of possible eruption points.

The geomorphic knowledge of behaviour of river system and its morphological characteristics viz. channel geometry, channel morphology and channel pattern, river metamorphosis, bank morphology etc. may help in controlling river floods through several flood control measures which include a series of steps to tame the menacing rivers such as :

(i) to delay the return of runoff resulting from torrential rainfall to the rivers,

(ii) to hasten the discharge of water (by straightening the meandering channels),

(iii) to divert the flow of water (through diversion channels),

(iv) to reduce the impact of floods (through contruction of protective embankments), and

(v) to forewarn the occurrence of floods. It may be mentioned that without the knowledge of nature of erosion in the upper catchment area and sediment load characteristics of the river the construction of levees to confine the flood water within the valley may prove disaster as if the rate of erosion is very high in the upper catchment area, resulting in high sediment load then there would be more sedimentation in the valley causing gradual rise in the river bed. This may lead to sudden flash flood whenever the levee is breached.

Earthquakes are both natural as well as man-induced geomorphic hazards. The geomorphic knowledge of the stability of terrain and probable impacts of man-made structures on crustal stability such as construction of dams and reservoirs is of paramount significance in identifying weaker zones which are likely to be affected by seismic events. Similarly, the geomorphic study of the nature of hillslopes and their associated lithologies enable us to know the stability or instability of hillslope and possible slope failures through landslides of various sorts. This knowledge would help in identifying and mapping of unstable hillslopes which may be avoided for human settlements and road construction.

Urbanization and Geomorphology : Geomorphic knowledge is of great significance in deciding the extent of urban growth in geomorphologically fragile zones such as unstable hillslopes with weak lihtology and frequent slope failures, alluvial riverine zones affected by frequent floods, areas of probable subsidence, coastal zones marked by active cliff recession, tectonically and seismically sensitive areas etc. The application of geomorphic knowledge in urban development comes under the domain of urban geomorphology. 'Urban geomorphology, a recent but more useful branch of applied geomorphology, is the study of landforms, and their related process, materials and hazards, ways that are beneficial to planning, development, and management of urbanized areas or areas where urban growth is expected' (Cooke, et. al, 1982).

'Thus, urban geomorphology (the description is reproduced from Savindra Singh et. al, 1988) deals with the lithological and topographic characteristics and geomorphic processes and hydrological conditions (in a city or town) which determine the size and rate of urbanization and stability of ground surface of urbanized localities. An urban geomorphologist has to play two-fold role i.e. the study prior to urban development (through field survey, terrain classification, identification and selection of alternative suitable locations for urban settlements meant for particular activities after proper studies of material composition and nature of geomorphic processes and palaeogeomorphology based on conventional and remote sensing techniques) and during and after urban development, the study of effects of natural events on urban community and the impact of urban development on natural environment and their remedial measures' (Savindra Singh, et. al, 1988).

'Various geomorphological problems hetherto not understood by the planners and engineers lead to destruction and damage to urban settlements in varying environmental realms viz. settling of foundation materials in the dry lands of oil-rich states and in the periglacial regions; destruction of foundations by weathering ,processes; damage of highways; damage to buildings through inundation during floods in the subtropical humid regions etc. 'All these and many other problems arise in part from mismanagement or misunderstanding of geomorphological conditions' (Cook, 1982). An urban geomorphologist is supposed to provide relevant information regarding the geomorphological conditions (ground surface conditions, sediment system in the river basin, movement of sand particles in the desert, weathering of underlying rocks and geomaterials, hydrology and palaeochannel conditions), the nature of geomorphic processes, mismanagement and faulty use of terrain and many more to the urban planners, environmental managers and engineers. Very little attention is paid towards understanding the geomorphological conditions before the development of existing urban centres mainly in the developing countries. With the result 'uncontrolled growth gives rise to squatter settlements or shanty towns as in many poor third world countries, and it commonly creates serious social and environmental problems' (Cook, et. al, 1982) (Savindra Singh, et. al, 1988).

Many of the alluvial cities located at the banks of the Ganga (viz. Allahabad, Varanasi, Patna etc.) and its tributaries are facing geoenvironmental problems because the city growth has taken place in otherwise unsuitable areas viz. lowlying flood plains (known as kachar). Allahabad city, located in the lap of two sacred rivers viz. the Ganga and the Yamuna, has witnessed its fast physical growth in the last 30 years in the outskirts except in the west in the geomorphologically unsuitable lowland terrain which is prone to recurrent floods, water logging, weak foundations, house collapse etc. The city authorities have miserably failed in checking the haphzard growth of houses in lowlying areas even in the 'kachar' land. They do not bother for geoenvironmental conditions of the existing city and the vacant land in and around it. Thus, the city will continue to reel under the hazardous problems of flood, waterlogging, village type (in Indian context) growth, defective drainage, increase in the pollutants to the drains and rivers. The regular raising of the Ganga and the Yamuna embankments no doubt temporarily protects the lowlying areas from floods but in long range it also helps in siltation and rise in river bed. How long the process of raising of embankment will continue ? The future physical growth of the city should be oriented towards geomorphologically sound areas of Jhusi and Phaphamau (Savindra Singh, et. al, 1988).

Various Engineering Works

Geomorphological techniques and knowledge become helpful in several engineering projects related to development activities e.g. construction of roads, dams and reservoirs (related to river valley projects), canals, laying of pipelines for petroleum and potable water, withdrawal of groundwater etc.

(A) Road construction is closely related to topographic features of the concerned area mainly land stability. The construction of road requires detailed and authentic information regarding geological structure, lithological and stratigraphic characteristics, nature of rocks, strength of surficial deposits, geomorphic history of the concerned area and porosity, permeability, joints, strength and compressibility of rocks lying under the ground surface etc. which no doubt come from geomorphic study of the area.

(1) The presence of permafrost in periglacial regions poses serious danger for road construction. The understanding of nature, behaviour and characteristics of permafrost and mechanism of periglacial processes is of paramount

significance for man if any development programme related to cultivation, settlement, highways, mining etc. is to be extended to permafrost areas because it poses greatest problems of construction and engineering works of all the other environmental processes. In fact, all aspects of human activities in periglacial areas are affected by permafrost and periglacial processes.

All of the constructions made by man for the colonization of Tundra region of North America (e.g. roads, railways, bridges, buildings, factories etc.) before the knowledge of deceptive nature of permafrost have been greatly deformed. It may be pointed out the active layer lying over permfrost thaws due to removal of vegetation leading to collapse of surface layer and formation of thermokarst lakes. Thus, the collapse of ground surface also allows the roads and rails to collapse. The laying out of rails for Trans-Siberian rail line in Siberia suffered great damages in the initial stages because of thawing of permafrost due to vegetation removal and consequent subsidence of ground surface. Since then the importance of applied geomorphology in permafrost areas has gained currency. Now greater attention is paid towards the study of permafrost in Canada, U.S.A. and Russia. The PISAS (Permafrost Institute of the Siberian Academy of Sciences at Yakutsk) of Russia and CARREL (Cold Region Research and Engineering Laboratories, at Hanover, New Hampshire) of the USA are actively engaged in the study of permafrost and periglacial processes. Geocryology, an independent branch of periglaciology, has been developed to study periglacial environment and periglacial landscapes. The areas having permafrost, when selected for the construction of roads and laying of rails are treated for required period of time by removing vegetation cover and exposing the surface to direct insolation. Hot-oil pipe lines cause thawing of permafrost which may result in environmental hazards but suitable engineering works may minimize the problem.

(2) The areas of carbonate rocks mainly limestones pose problems for road construction. A geomorphologist may provide relevant information about sub-terranean waterflow, caves underneath and other landscape characteristics of karst topography. If the roads or rails are constructed in the areas having hidden limestone features without prior knowledge, the roads and rails collapse and are damaged due to collapse of roofs of limestone caverns. Thus, the karstic surface having lapies, solution holes (sink and swallow holes, uvalas, polje, doline etc.), sinking creeks, blind valleys etc. is not favourable for surface transport network. It is, thus, obvious that the morphological maps may be useful for such engineering works.

(3) Glaciated areas are also not favourable for road and rail construction because

glaciated features mainly depositional one such as morainic ridges mainly terminal moraines, eskers, drumlins, rochemouttonee (erosional) etc. present difficulties in road construction as more additional work is required for the purpose of road construction in glaciated areas where till plain, being more or less flat and rolling plain, is more suitable for road construction.

(4) Road construction requires detailed knowledge of stratigraphy of the concerned area. The regions having buried lakes, buried palaeochannels, aquifers with high watertables, man-infilled depressions etc. are unfavourable for road construction. The areas having lacustrine deposits should be avoided.

(5) The knowledge of slope stability, safety factor, history of slope failure, lithological characteristics etc. is very important for road construction in mountainous areas with steep hillslopes. Since the construction of roads along hillslope requires a lot of excavation work and hence if the hillslope is fragile and prone to slope failures i.e. frequent landslides, mud flows, earth flows etc., the construction of roads further accentuates slope failure, triggers massmovement and presents high risk factor for road transport. Such information may be available through detailed geomorphological investigation of the concerned area.

(B) Dam construction, besides being related to engineering work, is also concerned with geology, geophysics, seismology and geomorphology and requires the information about the geological structure and stratigraphy characteristics, seismic resistivity, rock resistance, erodibilty of geomaterials, porosity and permeability of rocks at the proposed site of dam construction (to assess loss of water to underground region through seepage and percolation from the base of reservoir), safety and risk factor of hillslope and valleyside slope, nature of transverse section of the valley, nature of longitudinal profile of the concerned river, channel gradient, relative hardness of rocks of the proposed site, rainfall in the catchment area, nature and amount of runoff of the catchment area above the dam, volume, velocity and discharge of the river, sediment load of the uppercatchment, probable loss of water from the reservoir through evaporation etc. Most of these information may be available through geomorphological study of the concerned locality. The morphometric properties of the drainage basin, analysis of hydraulic geometry, channel morphology and hydrodynamics provide useful information about the drainage basins which may be helpful in selecting suitable sites for the construction of dams. The determination of hierarchical orders of the fluvially originated drainage basin provides detailed information about the number of stream segments, catchment areas, water discharge etc. at each confluence point. This helps in determining the volume of water, discharge of water, flow velocity, sediment load and sediment transport, sedimentation etc. at specific point of a river which facilitate dam construction. Generally, dam is constructed at the point across the river where :

(i) the river valley is very narrow and constricted so that the length of dam is shortened to maximum possible extent,

(ii) rocks are resistant and least erodible,

(iii) catchment area above the proposed dam site is quite extensive so that required volume of water (runoff) may be available,

(iv) sediment load of the river is low so that the reservoir may not be silted,

(v) the reservoir floor is of impermeable rocks so that there is minimum loss of water through seepage and percolation etc.

(C) Construction of air strips for landing and takeoff of aeroplanes require high engineering skill and thus engineers while selecting the sites for the construction of air strips (aerodrome) may be benefitted from geomorphologists because the landscape characteristics determine the sites for air strips which are generally constructed where (i) extensive flat terrain surface with resistant geomaterials (including unconsolidated surficial deposits) is available so that safe and suitable runway may be built, (ii) ground surface has almost level slope, (iii) the area is free from flood, (iv) there is absence or low frequency and intensity of fog, (v) there is maximum visibility etc. The relevant information regarding the quality and characteristics of terrain may be available from morphological map of the area concerned.

Hydrology and Geomorphology : The landscape and terrain characteristics, besides geology, are closely related to the occurrence, availability, quality and utilization of usable water for different purposes by man from different sources e.g. surface water (stream water, lake water, spring water), groundwater etc. Different stratigraphic and lithological zones present different conditions of surface and groundwater. There is direct relationship between limestone terrain and surface and groundwater. Carbonate rocks mainly limestones are porous and permeable. This porosity and permeability develops in two ways-(i) primary porosity and permeability develops with the formation of joints and cavities at the time of formation of limestones, and (ii) secondary porosity is caused due to earth movements (i.e. faulting, folding, tilting, warping etc.) and solution (corrosion) mechanism. In the initial stage of karst cycle of erosion there is abundance of surface water and surface drainage due to enough surface runoff but with the advancement of karst cycle of erosion solution process becomes more active resulting into the development of numerous solution holes (e.g. sink and swallow holes) and hence surface water disappears through these solution holes to become sub-terranean (underground) water with the result most of surface drainage disappears. This mechanism results in marked reduction in surface water and the local people have to depend on karst spring water which may be useful or unsuitable for human use.

The groundwater is contaminated due to infiltration of polluted water, during rainfall, through solution holes because the water is not properly filtered through geomaterials as the holes allow quick downward movement of polluted water mixed with pollutants and bacteria. Such water is injurious for health and hence should not be used as potable water. The location of such holes, which allow downward movement of polluted water, can be easily determined through the use of colour. The geomorphic information of limestone areas also helps in digging of dug wells. If the limestones have enough porosity and are capped by sandstone layer, there is every possibility for rich groundwater reserve and the water is also naturally filtered while passing through sandstone beds and hence groundwater is free from pollutants. Limestone areas are not suitable for dam and reservoir construction because it requires a lot of expensive engineering work.

The geomorphic history of glaciated landscape, characteristics of different types of glacial deposits and landform characteristics help in searching and determining water potential of glaciated regions. Outwash plains, valleytrains, intertill deposits etc. are supposed to contain abundant groundwater. True till deposits are poor in retaining groundwater because they do not form ideal aquifers. Buried periglacial and interglacial valleys under surficial ground material provide large and rich groundwater reserves. This situation of buried periglacial valleys may be known only through geomorphological study of concerned area i.e. denudation chronology and geomorphic history of the area concerned help in this regard. Aquifers with sand formation contain large groundwater reserves as is the case of the Ganga plain of North India.

Exploring Minerals

There is a close relationship between mining geology and geomorphic history of the concerned area. It is well known fact that mineral deposits are closely associated with geological structure. Such structures are ascertained on the basis of landscape characteristics of the specific locality. Some minerals are associated with specific landscapes. Thus, denudation chronology of a particular locality provides all those information related to landscape evolution and climatic changes under which particular mineral was formed.

(1) Some of the minerals are identified and located on the basis of their specific topographic expressions. For example, quartz formed in veins of surface rocks are found in the form of ridges on the ground surface. Similarly, lead-zinc veins are also found in the form of ridges. On the other hand, some minerals have negative topographic expressions (lowlands, depressions, troughs etc.).

(2) Weathering is also closely related to minerals, because many of the minerals are weathering residues like some iron ore, clay mineral, caliche, bauxite, manganese, nickel etc. Denudation history, thus, helps in locating such residue minerals, which are mostly found on primary erosion surfaces. It may be mentioned that minerals change their places during youth and mature stages of cycle of erosion but they become stabilized in the penultimate stage of terrain development and remain at their places on eorsion surfaces. Bauxite minerals are associated with weathered basaltic laterite surfaces of the patlands of the Ranchi plateau and Palamau upland (Bihar) e.g. Bagrupat has rich bauxite reserve. It may be mentioned that eluviation (downward movement) of iron and aluminium content through percolating rainwater through basaltic cover and illuviation of these materials below upper surface have formed strong bauxite cuirass (alcrete hard pan) on 'pats' (basaltic-lateritic capped mesas and buttes). Thus, these pats are clearcut expressions of bauxite. Laterites also contain residues of manganese. Many more examples may be cited.

(3) Placer deposits are mixtures of heavy metals which represent aggregates of materials derived through chemical weathering and erosion of metallic formations e.g. gold, diamond, silver etc. Nine placer deposits, which contain precious minerals, have been identified so far viz. aeolian placer, alluvial placer, residual placer, bajada placer, beach placer, glacial placer, buried placer, ancient placer etc. Placer, in fact, means a superficial deposits by geomorphic agents from which precious minerals such as gold, silver etc. are washed. Residual placer is formed at situ due to weathering of rocks but such placer is economically not significant. On the other hand, when weathering residues are transported downslope and are deposited in the form of colluvium, these are called colluvium placers which generally contain gold. Alluvial placers are formed due to deposition of weathering and erosion residues by fluvial process. Such alluvial placers contain most of world reserve of platinum. Besides, alluvial placers also contain gold, tin, diamond etc. Bajada placers are generally found at the foothills in the arid and semi-arid areas. Beach placers contain different precious minerals such as gold, diamond, zircon etc. It is, thus, apparent that the study of depositional features of different geomorphic features in varying environmental conditions may help in locating placer deposits which contain precious metals.

(4) Geomorphic principles help in the exploration of mineral oil. Generally, mineral oil is found in the porous and permeable lithologies having lower layers of impermeable rocks. Sandstones and limestones provide ideal locations of mineral oil because of their high porosity and permeability while shale below sandstones or limestones acts as impermeable layer which does not allow downward movement

of mineral oil. There are two concepts of mineral oil formation viz. organic concept and inorganic concept. According to organic theory oil is formed due to decay and decomposition of organic matter. After formation, mineral oil needs storage in the rocks wherein mineral oil is trapped under two situations viz. under structural traps, and stratigraphic traps. Sedimentary strata are folded into anticlines and synclines which allow permeable and impermeable strata to come closer so that mineral oil is well confined and preserved within upper permeable and lower impermeable beds. Thus, folded structure of sedimentary formation is most probable condition for the occurrence of mineral oil wherein oil, when drilled, oozes out due to hydrostatic pressure.

There are many more areas where geomorphological knowledge may be applied e.g. land use management, environmental management, natural disaster reduction, soil erosion and sedimentation, watershed management, river management, coastal zone management etc. The techniques of morphological mapping, terrain classification and evaluation, air-photo interpretation, remote sensing techniques, process measurement etc. may be fruitfully used in the areas which are concerned with development process and overall socio-economic growth of society.

Bibliography

Agnew, J.A. : *The City in Cultural Context,* Allen & Unwin, New York, 1984.

Aitchison, J.W. : *The Farming Systems of Wales : A Study of Spatial Variability,* University College of Wales, Department of Geography, 1970.

Alexander, B. : *Geography : Realms, Regions, and Concepts,* John Wiley & Sons, New York, 2000.

—— : *Human Geography : Culture, Society and Space,* John Wiley & Sons, New York, 1999.

—— : *The Earth : An Introduction to its Physical and Human Geography,* John Wiley & Sons, New York, 1995.

Alexander, J.W. and Gibson, L. James : *Economic Geography,* Prentice Hall, New Jersey 1979.

Arvill, R. : *Man and Environment Crisis and the Strategy of Choice,* Penguin, New York, 1967.

Balchin, W.G.V. : *Geography : An Outline for the Intending Students,* Routledge & Kegan Paul, London, 1970.

Bansil, P.C. : *Agricultural Problems of India,* Vikas Publishing House, New Delhi, 1977.

Bansil, P.C. : *Agricultural Statistics in India,* Arnold Heineman, New Delhi, 1974.

Bater, J. H. : *Russia and the Post-Soviet Scene : A Geographical Perspective,* New John Wiley & Sons, New York, 1996.

Bennett, D. : *World Population Problems,* Park Press, Campaign, 1984.

Berger, W.H. : *Abrupt Climatic Change,* D. Radiel Dordrecht, Boston, 1990.

Bhatia, B.M. : *Poverty, Agriculture and Economic Growth,* Vikas Publishing House, New Delhi, 1977.

Bose, A. : *India's Urbanization,* Kanishka Pub., New Delhi, 1980.

Brian, J.L. : *Geography of Market Centers and Retail Distribution,* Prentice-Hall, New Jersey, 1967.

—— : *The Human Consequences of Urbanization,* St. Martin's Press, New York, 1973.

Brun, S.D. : *Cities of the World : World Regional Urban Development*, Harper & Row, New York, 1983.

Brush, John E. : *The Morphology of Indian Cities*, Popular Prakashan, Bombay, 1962.

Burton, Robert, W.K. and Gilbert, F.W. : *The Environment as Hazard*, Oxford University Press, New York, 1978.

Calder, C. C : *An Outline of the Vegetation of India: An Outline of the Field Sciences of India*, Indian Science Congress Association, Calcutta, 1937.

Carlson, L. : *Geography and World Politics*, Prentice Hall, Englewood, 1958.

Chandna, R. C : *A Geography of Population*, Kalyani Publishers, Ludhiana, 1992.

Chauhan, V. S. and Gautam, A. : *Comprehensive Geography of India*, Rastogi Publications, Meerut, 1996-97.

Clark, W.A.V. : *Migration and Development*, The Hague, Mouton, 1975.

Clarke, J.I. : *Geography and Population : Approaches and Applications*, Pergamon Press, New York, 1984.

Claval, Paul : *Introduction to Regional Geography*, Blackwell, Maiden, 1998.

Cole, M.M. : *The Savannas : Biogeography and Geobotany*, Academy Press, London, 1986.

Cressey, G.B. : *Asia's Land and Peoples*, McGraw-Hill, New York, 1963.

Crother, G. : *India, A Travel Survival Kit*, Berkley, U.S.A.,1990.

Datt Ruddar and Sundharam, K. P. M. : *Indian Economy*, S. Chand & Company Ltd., New Delhi, 1991.

Deshpande, C. D. : *India : A Regional Interpretation*, Indian Council of Social Science Research, New Delhi, 1992.

—— : *India : A Regional Interpretation*, Northern Book Centre, New Delhi, 1972.

Detwyler, Thomas R. : *Man's Impact on the Environment*, McGraw-Hill, New York, 1988.

Dev, S. M. and Ranade, A. : *Persisting Poverty and Social Insecurity : A Selective Assessment*, Oxford University Press, New Delhi, 1999.

Dickinson, R.E. : *The Regional Concept*, Routledge Kegan Paul, London, 1976.

Dikshit, R.D. : *Geography and the Teaching of the Environment*, Department of Geography, Poona, University, 1984.

Dubey, R. N. : *Economic Geography of India*, Kitab Mahal, Allahabad, 1998.

Edgar, A. : *Plants, Man and Life*, University of California Press, California, 1967.

Ehrilich, P.R. : *Eco-Science*, W.H. Freeman & Co., San Franciscess., 1977.

Eisner, W.B. : *The Middle East : A Physical, Social and Regional Geography*, Methuen, New York, 1978.

Enyedi, Gyorgy : *Natural Geography,* Applied Geography in Hungary, Budapest, 1964.

Farmer, B.H. : *Green Revolution,* MacMillan, London, 1977.

Febure, L. : *Geographical Introduction to History,* MacMillan, London, 1925.

Fitz Gerald Brian P. : *Science in Geography,* Oxford University Press, London, 1974.

Futehally, I. : *Our Environment,* N.B.T., New Delhi, 1988.

Gansser, A. : *Geology of the Himalayas,* John Wiley & Sons, London, 1964.

Gilbert, Alan, : *The Meha City in Latin America,* United Nations Library Press, New York, 1996.

Gill, K. S. : *Evolution of the Indian Economy,* N.C.E.R.T., New Delhi, 1993.

Goudie, A. : *The Human Impact on the Natural Environment,* Basil Blackwell, Oxford, 1986.

Gregory, D. : *Ideology, Science and Human Geography,* Hutchinson, London, 1978.

Grigg, D.B. : *Geographical Impact of Migration,* White and Woods, Longman, London, 1980.

Haggett, P.: *Network Analysis in Geography,* Edward Arnold, London, 1969.

Haq, M.U. : *Human Development in Changing World,* Oxford, New York, 1992,

Hathaway, D.E. : *Government and Agriculture : Public Policy in a Democratic Society,* MacMillan, London, 1963.

Herskovits, M.F. : *Cultural Anthropology,* Alfred A. Knopf, New York, 1955.

Husain, M. : *Agricultural Geography,* Inter India Publications, New Delhi, 1979.

—— : *Evolution of Geographical Thought,* Rawat, Jaipur, 1995.

—— : *Mapping of Environment and Sustainable Development,* Jamia, New Delhi, 1996.

—— : *Nagaland : Habitat, Society and Shifting Cultivation,* Rima Publishing House, N7ew Delhi, 1990.

—— : *The Valley of Kashmir,* Rajesh Pub., Delhi, 1998.

Isard, W. : *Location and Space Economy,* MIT Press, Massachusetts, 1965.

Jacobson, J.L. : *Planning the Global Family,* World Watch Institute, Washington, 1987.

Jhingran, A. G. : *Geology of the Himalayas,* Oxford University Press, New Delhi, 1981.

Johnson, B. L. C : *India : Resources and Development,* Arnold -Heinemann, London, 1980.

—— : *South Asia,* Heinemann Educational Books Ltd., London, 1969.

Johnson, B.L. : *India : Resources and Development,* Barnes & Noble, New Jersey, 1979.

Johnston, R.J. : *Philosophy and Human Geography,* Edward Arnold, London, 1983.

—— : *A Dictionary of Human Geography,* Black-well, Oxford, 1990.

—— : *City and Society : An Outline for Urban Geography,* Hutchinson, London, 1984.

Joshi, H. L. : *Industrial Geography of India,* Rawat Publications, Jaipur, 1990.

Kosinski, L.A. : *People on the Move,* Methuen, London, 1975.

Kostabade J.T. : *Essentials of World Regional Geography,* Sunders College Publishing, New York, 1992.

Kumar, A. : *Chotanagpur Highlands : A Study in Synchroniety,* Thinker's Library, Allahabad, 1985.

Kumar, P. : *Madhya Pradesh : Ek Bhaugolik Adhyayan,* (Hindi), Madhya Pradesh Hindi Granth Academy, Bhopal, 1996.

Lamb, A. : *The China-India Border : The Origin of the Disputed Boundaries,* Oxford University Press, New York, 1964.

Lewin, K. : *Field Theory in Social Science,* University of Chicago Press, Chicago, 1951.

Lewis, G.J. : *Human Migration,* MacMillan, London, 1982.

Mackinder, Halford, J. : *Democratic Ideals and Reality,* Holt, New York, 1919.

Majumdar, D. N. : *Races and Cultures of India,* Asia Publishing House, Bombay, 1958.

Mannion, A.M. : *Global Environmental Change,* Longman, New York, 1991.

Marshall, B. : *The Real World,* Houghton Miffilin Company, Boston, 1991.

Mash, G. P. : *Man and Nature Physical Geography as Modified by Human Action,* Charles Scriber, New York, 1964.

Maxwell, K.E. : *Environment of Life,* Dickenson Publishers, Co., California, 1973.

Michael, C. : *Rural Settlement and Land Use : An Essay in Location,* Hutchinson, London, 1966.

Misra, V. C : *India-Regional Studies,* 21st IGU Publication, New Delhi, 1968.

Miller, G.T. : *Living in the Environment,* Wadsworth Press, Belmont, Calif., 1994.

Misra, J. P. : *Bharat Ka Bhoogol* Geography of India, Vidya Sagar, Allahabad, 1997.

Misra, R.P. : *Development Issues of our Time,* Concept, New Delhi, 1984.

—— : *National Development,* Vikas Publication, New Delhi, 1965.

—— : *Regional Planning : Concept, Technique and Policies,* Concept, New Delhi, 1992.

Muller, S.H. : *The World's Living Languages,* Oxford, New York, 1964.

Myrdal, Gunnar : *Rich Lands and Poor,* Harper & Brothers, New York, 1957.

Nielsen, N.C. : *Religions of the World,* Oxford, New York, 1983.

Ojha, S. S. : *Geography of India,* Bhaugolik Adhyayan Sansthan, Allahabad, 2001.

Pannikar, K. M. : *A Survey of Indian History,* Popular Prakashan, Bombay, 1964.

Paul, A. : *World Regional Geography : A Question of Place,* Harper & Row, New York, 1977.

Pedelaborde, P. : *The Monsoon,* Methuen and Co., London, 1963.

Perpillou, A.M. : *Human Geography,* Longman, New York, 1977.

Persson, G.A. : *Resources and World Development,* John Wiley, Chichester, 1987.

Prescott, j.R.V. : *Political Frontiers and Boundaries,* Allen & Unwin, Winchester, 1987.

Qureshi, M. N. : *Himalayan Geology,* Oxford, New Delhi, 1971.

Rao, B.P. : *Bharat : Ek Bhaugolik Samiksha,* Vasundhara Prakashan,Gorakhpur, 1995.

Rapp, A. : *Human Activity and Environmental Processes,* John Wiley, Chichester, 1987.

Ray, S. : *Geomorphology of India,* Firma, K. L. M. Pvt. Ltd., Calcutta, 1978.

Rubenstein, J.M. : *The Cultural Landscape,* Prentice Hall, London, 1983.

Sen, Bandhudas : *The Green Revolution in India : A Perspective,* Wiley Eastern, New Delhi, 1974.

Sen, L. K. : *Planning Rural Growth Centres for Integrated Area Development : A Study in Miryalguda Taluka,* N. I. C. D., Hyderabad, 1971.

Shafi, M. : *Ganga-Yamuna Doab, in India : Regional Studies,* National Committee for Geography, Calcutta, 1923.

——: *Agricultural Productivity and Regional Imbalances—A Study of Uttar Pradesh,* Concept Pub. Co. : New Delhi, 1983.

Sharma R. C : *Readings in General Geography and Geography of India,* Jawahar Publishers, New Delhi, 1992.

Sharma, H. S. : *Environmental Design and Development,* Scientific Pub., Jodhpur, 1988.

Sharma, R.C. : *Settlement Geography of Western Rajasthan,* Rawat Pub., New Delhi, 1972.

Shaw, J.B. : *The Post Soviet Republics : A Systematic Geography,* John Wiley & Sons, New York, 1992.

Shiva, V. : *The Violence of Green Cultivation,* Mapusa, Goa, 1991.

Short, J.R. : *Introduction to Urban Geography,* Routledge & Kegan Paul, London, 1984.

Singh, H. H. : *Geography and Environment,* Concept, New Delhi, 1986.

Singh, J. : *An Agricultural Geography of Haryana,* Vishal Publication, Kurukshtra, 1976.

Singh, L. R. : *The Tarai Region of U.P. : A Study in Human Geography,* Ram Narain Beni Prasad, Allahabad, 1965.

Small, R. J. : *The Study of Landforms,* Cambridge University Press, London, 1970.

Smith, D.M. : *Human Geography : A Welfare Approach,* Edward Arnold, London, 1977.

Srinivasan. T.N. : *The Green Revolution,* Indian Statistical Institute, Calcutta, 1971.

Stamp, L.D. : *Applied Geography*, Penguin, Harmondsworth, 1960.

Sundaram, K. V. : *Geography of Under-development : The Spatial Dynamics of Under-development*, Concept, New Delhi, 1983.

Tarrant, J.R. : *Agricultural Geography* Newton Abbot : David and Charles, New York, 1974.

Taylor, James A. : *Weather and Agriculture*, Pergamon, Oxford, 1967.

Taylor, P.J. : *Geography : Its Nature*, MacMillan, London, 1986.

Turner, Roy : *India's Urban Future*, University of California Press, Los Angeles, 1962.

Vayda, A. P. G. : *Man in the Pacific Islands : Essays on Geographical Change in the Pacific*, Oxford University Press, New York, 1972.

Verma, R. V., *Bharat Ka Bhaugolik Vivechan* Hindi, Kitabghar, Kanpur, 1956.

Weyer, E.M. : *The Eskimos : Their Environment and Folkways*, Yale, 1932.

Willens, A. : *Dry Lands, Man and Plants*, Architectural Press, London,

Wirth, Louis : *On Cities and Social Life*, University Press, Chicago, 1964.

Yantis : *The Encyclopaedia of World Cultures-Africa*, MacMillan, New York, 1993.

Zelinsky, W. : *Geography and a Crowding World*, Oxford University Press, New York, 1970.

Zimmermann, E.W. : *World Resources and Industries*, Harper and Row, London, 1951.

Index

E

F

G

❑❑❑